PENGUIN BOOKS

Nuclear Free — The New Zealand Way

David Lange was born in Otahuhu, New Zealand, in 1942. He practised law in Auckland until 1977, when he was elected member of Parliament for Mangere. In 1983 he became Leader of the Opposition, and then Prime Minister the following year. He resigned from the office of Prime Minister in 1989.

David Lange remains active in politics, speaking and writing at home and abroad, often on behalf of the nuclear-free policy which was a hallmark of his Labour Government.

Nuclear Free
The New Zealand Way

David Lange

PENGUIN BOOKS

PENGUIN BOOKS

Penguin Books (NZ) Ltd, 182–190 Wairau Road, Auckland 10, New Zealand
Penguin Books Ltd, 27 Wrights Lane, London W8 5TZ, England
Penguin USA, 375 Hudson Street, New York, NY 10014, United States
Penguin Books Australia Ltd, 487 Maroondah Highway, Ringwood, Australia 3134
Penguin Books Canada Ltd, 10 Alcorn Avenue, Toronto, Ontario, Canada M4V 1E4

Penguin Books Ltd, Registered Offices: Harmondsworth, Middlesex, England

First published 1990
1 3 5 7 9 10 8 6 4 2

Editorial services by Michael Gifkins and Associates
Designed by Richard King
Typeset by Typocrafters Ltd, Auckland
Printed in Australia

CONTENTS

INTRODUCTION

The *Britannia* Stayed Away

EARLY IN 1990, Queen Elizabeth II arrived in New Zealand. She came to celebrate the one hundred and fiftieth anniversary of the signing of the treaty between the Crown and the Maori people by which New Zealand became a British territory. There was some disappointment that she was not accompanied by the Royal Yacht *Britannia*. This distinctive vessel was not glimpsed among the many that took part in the water-borne celebrations of the signing of the treaty. *Britannia*, once a welcome sight in New Zealand waters, was no more seen here.

RY *Britannia* was a vessel of the Royal Navy. In time of peace it was a pleasure craft. In time of war it was a hospital ship. As a naval vessel, it was the property of the British Government. That government would not allow its property to come to New Zealand. If the ship appeared in New Zealand waters, the British Government could no longer confound its enemies with speculation that RY *Britannia* was armed to the gunwales with nuclear weapons.

The capacity of RY *Britannia* to deal death and destruction was brought into question when a new law came intc force in New Zealand. The law held that foreign warships (of which, technically, *Britannia* was one) would be refused entry to New Zealand waters unless the Prime Minister was satisfied that the ships were not carrying nuclear weapons. When I was Prime Minister I was willing to pronounce myself perfectly satisfied that RY *Britannia* was innocent of nuclear weaponry. It didn't glow in the dark, other than when illuminated, and its crew were mostly waiters and bandsmen. *Jane's Fighting Ships* was silent on its aggressive capabilities. I was once laid low by a prawn I ate on board, but I thought that was the limit of its lethal powers. I was glad to welcome *Britannia* to New Zealand.

The *Britannia* stayed away. It was the policy of the British

Government neither to confirm nor deny whether the ships of the Royal Navy were armed with nuclear weapons. That government felt that sending *Britannia* to nuclear-free New Zealand would amount to a denial that the Royal Yacht was carrying nuclear weapons. *Britannia*, and its battery of butter knives and cocktail forks, stayed, its capacity for menace uncompromised, in its home port.

In one way, I can't be sorry that *Britannia* never made it to the treaty's birthday party. Senior members of the British Government (like senior members of other Western governments) often took it on themselves to tell me how misguided was New Zealand in making itself nuclear-free. When the same people refused to acknowledge in public that the Royal Yacht was not carrying nuclear weapons, I knew it wasn't me who'd taken leave of my senses.

Many of the assumptions behind nuclear deterrence are, when you take them to pieces, as nonsensical as the posturing which kept *Britannia* out of New Zealand. This has nothing to do with the weapons themselves. Their destructive power is unprecedented, but in the end they're no more than machinery. They could all be disarmed tomorrow if we had the will to do it. It's the will that matters. The answer to nuclear weapons lies in the reasons why governments put themselves in thrall to doctrines which essentially have it that nations can protect themselves by threatening to blow up the planet. It lies in knowing what breaks the spell that keeps governments from seeing that reliance on nuclear weapons is worse than foolish.

This book is essentially the story of how one small country found the political will to say no to nuclear weapons. It tells how small-scale protest eventually turned into popular movement and then to political action. It tells how government and public joined together to put nuclear-free New Zealand beyond ordinary politics.

In its way, it's a story of the best kind of politics, the kind of politics that points the way to a better future and battles through until it gets there. Most politics in New Zealand, I have to say from experience, are as dismal as politics anywhere else. The politics of nuclear-free New Zealand were always very different. I was in a government with a mandate from the electorate to take substantial steps to reduce the dangers to New Zealand of nuclear weapons. In putting that mandate into practice, I ran into many difficulties. Some were traps set for me by others. Some were totally self-inflicted. For all the problems I faced and all the false starts I made, I knew it was worth it. It felt right. I always counted on the support of the people who elected the government, and eventually, many who didn't. In the end, it was the ordinary people of New Zealand who made their country nuclear-free, and it is the same people who can make sure it stays that way.

This book is a personal view of what happened. It doesn't have any claim to objectivity. It's an eye-witness account, although hindsight inevitably puts what I saw at the time in a new light. Even the act of writing down the events that took place can give them a coherence they didn't originally have. Much of what's described here seemed like a shambles when it happened. It would have been good when I was Prime Minister to have the luxury of thinking only about our foreign policy, but that's not the way it works. If some of the confusion comes through in the text, that's probably what it was like.

Given its viewpoint, the book doesn't try to explain what happened in the Pentagon or the State Department, or in the minds of the people who own the *Britannia*. I don't try to account here for the actions of others. I've limited myself to what I thought about them, or the effect their actions had on me. Nor can this book do more than touch on the wider history of the nuclear-free movement in New Zealand. It happened that I took part in the very early days of anti-nuclear protest here. I was again, less directly, involved when the campaign for nuclear-free New Zealand became a more popular movement in the 1970s. But most of the book deals with my time as a member of Parliament and the difficult path of the nuclear-free policy from idea to reality. It ends with nuclear-free New Zealand beyond my power as a politician to influence; I think beyond the power of any politician ever to influence. And that, I believe, is exactly where it should be.

1

A Vision of Apocalypse

IN THE SKY above Otahuhu the moon turned to blood.

Otahuhu in 1962 was a township half an hour's drive from the city of Auckland where, a law student not yet twenty years old, I lived with my parents, brother and sisters, brought up on the extraordinary prophecies of the *Book of Revelation* but not expecting their fulfilment in my lifetime. As I walked from the bus stop to my home on a cold winter evening I saw an eerie lightening of the sky. Beams of light radiated from the northern horizon and intersected with each other through the blackness of the night. I climbed to the open balcony on the first floor of the old house and was arrested with awe as the sky pulsated with these brilliant shafts of light. They were red and white. They extended across the night like the ribs of a fan. They were spinning, they were intermingling. The sky was diffused with a ghastly brush of red. It was an unnerving spectacle.

For all that I told myself it was unlikely that divine judgment was to be visited on the people of Otahuhu, I couldn't reason away the chill sweat of dread. I had seen in the South Island of New Zealand the majesty of the aurora australis. What I saw that night came from the wrong pole. The intensity of light and colour was far more extreme. But as the minutes passed, it dawned on me that I had not been promoted to glory. No fundamentalist rapture had taken place. I was still affected by gravity and composure crept back. I reverted to my usual dismissal of divine phenomena. After about a quarter of an hour the lights died away.

Eventually the radio told us what had happened. Thousands of miles north-east of New Zealand, the United States had tested a one-megaton nuclear weapon at high altitude by launching it on a rocket and exploding it two hundred miles above Johnston Island in the Pacific. The device had been laced with millions of copper threads. They danced in the ionosphere and ruptured the ordered magnetic fields, which, in spasmodic correction of the violation, created a nuclear aurora.

For me the technical explanation made things worse. The confusion in the sky that night haunted me as a vision of a man-made apocalypse, a terrifying retaliation of natural forces against the evil of unnatural invasion and a warning that a small country at the edge of the world in the South Pacific was no longer far enough away from the quarrels of the great powers to escape their consequences. It was a shock to realise that the power of nuclear weapons could straddle the world and unleash a threat on an inoffensive country like New Zealand.

New Zealand in 1962 was a place apart. My father, a doctor with a passing interest in novelties and a watcher of planes landing at the local airfield, hadn't bought a television set. There wasn't a lot of point. When television arrived there was one channel in the evenings only, in grainy black and white. There were no private radio stations. The government owned and ran a commercial and a non-commercial radio network. The bars closed at six in the evening and stayed closed all day on Sunday. Every session of the cinema began with a grand rendition of the British national anthem played by a British military band on the soundtrack of a film that showed the Queen riding side-saddle in military attire at the Trooping of the Colour in London. Unquestioningly the audience stood when the first drumroll of the anthem sounded.

Just in front of our family house was a small triangular park containing two monuments. The first, an obelisk of sandstone, commemorated a man who died a century before, twelve thousand miles from his English home, trying to disposess the native Maori people of their property. The second monument was the war memorial. A magnificently cast horse bearing a fully kitted-out cavalryman was mounted on a large concrete plinth on the sides of which were bronze plaques recording the sacrifices made between 1914 and 1918. Once a year, on 25 April, the people of Otahuhu gathered around the war memorial for the Anzac Day service.

More sacred than any Sunday, Anzac Day was named for the

Australia and New Zealand Army Corps and marked the slaughter these men were consigned to at Gallipoli in 1915. The war they fought in not having ended all wars, the day of their tragedy became a memorial to all who died while serving New Zealand abroad.

My earliest solemn memories are of the Anzac Day services at Otahuhu. The main road south from Auckland, the Great South Road, was closed, and the streets filled with the townsfolk. The veterans marched on. Schoolteachers and the local grocer and barber looked very different one day a year, bedecked with medals. The Salvation Army band led the hymn singing. The mayor intoned a remembrance. The bugler played the Last Post. The guard of honour from the local school cadet unit fired three volleys of blanks. The less sensitive of the children ran through the ranks gathering the spent cartridges. I only did it once; my father thought it poor form. Reveille sounded and the wreath-laying began.

This part of the ceremony was hard to take. It was not the official tributes that churned me up — it was seeing the mothers of school friends placing homegrown bunches of flowers against the base of the memorial in memory of the fathers my friends had never seen. Each year I saw an old woman place two blooms against the monument, one for her brother killed at Gallipoli and the other for her son killed nearly thirty years later in Egypt.

There were hundreds of people at the Anzac Day service. They all knew each other. I knew most of them and they knew me because everyone knew the doctor's family. It was a good time and place to grow up in. Although I was the doctor's son and we lived in a big house, I went to the same schools as everyone else, talked like everyone else and felt at home in any house in Otahuhu. It was that sort of a town. The plumber probably earned as much as my father. Otahuhu had no unemployment. Hardly anyone was out of work in New Zealand. It was a rich little country and there didn't seem to be an awful lot to worry about.

Otahuhu was no place for a nuclear aurora and its intrusion shattered my complacency. The nuclear afterglow persuaded me that my country was not a Noah's ark. There would be no lucky survivors and no shelters deep enough if the world went mad. I was struck by the conviction that the planet's future depended on our eliminating any possibility of nuclear war.

Having reached this conclusion, I was unsure how to act on it. New Zealand had a military alliance with the United States, a nuclear power, but nuclear weapons played no immediate part in our defence. We were always a bystander. Opposition to nuclear testing could only be a show of feeling. There wasn't a clear target for our resistance. Our impotence was plain when, in the year the nuclear aurora shone on Otahuhu, France announced that it was moving its atmospheric nuclear weapons testing programme from the Sahara Desert to its island territories in the South Pacific. My concepts of distance were not then tempered by the experience of travel, and I assumed that testing on the atolls of Tahiti was testing next door to New Zealand. My new-found fear of nuclear destruction was fuelled by xenophobia. The South Pacific was ours, not a laboratory for the powers of the Northern Hemisphere. I remember self-righteously feeling how extraordinarily grovelling Australia was to allow nuclear weapons to be tested on their soil by the British. What the French, and the Americans, were doing in the Pacific seemed to me to be an affront to the independence of New Zealand.

I didn't join the Campaign for Nuclear Disarmament or any other group formed to protest against nuclear testing. It was through my law studies that I was drawn into the small active protest movement. I worked part-time through most of my law degree and in 1962 I got a clerk's job with Haigh, Charters and Carthy, barristers and solicitors of Auckland. In the early morning I went to lectures at Auckland University, worked at the office through the day and went to more lectures in the evening. Haigh, Charters was not a wealthy firm. The office accommodation was dreadful. It had the austerity of a Noncomformist chapel. The workload of the practice was no more fashionable. The firm acted for most of Auckland's trade unions. It was heavily involved in the legal defence of political activists. It was the legal comforter of the artistic and literary community. Haigh, Charters was an exciting firm to work for if you didn't want to build yourself a reputation among the moneyed classes.

Frank Haigh, the senior partner, was one of Auckland's leading advocates. He represented the maimed, the defamed, the dissenting and the dispossessed. Les Charters was the second senior partner. He was a great lawyer. Disqualified from practice and imprisoned for some years during the war because of his conscientious objection to military service, he declined to take the oath of allegiance to allow him to practise law again until 1958, when he was sure he was no longer likely to be called up. Jack Carthy, a devout Catholic, made up the trio. He was the general practitioner of the team. The agnostic Haigh was highly catholic in his championing of the right to pursue a cause. He acted for a clutch of Christian pacifists, for the Church of Scientology when it was the subject of a commission of inquiry, and for the Communist Party of New Zealand. Of them all, the last stood lowest in popular opinion.

The Communist Party carried a burden of contempt in New Zealand as elsewhere in the days of the Cold War. To public acclaim the National Government in 1951 smashed the waterfront union. Some of its influential members were communists; the odium lingered and politicians raised the bogey from time to time. Most of the party's few members were sure they were being spied on. Afraid that their party would be outlawed and its assets confiscated, they spent no small part of their income in devising trusts and other legal shelters which would protect their property when the government finally lived up to its threats and the axe fell on them.

Under the guise of many names, the Communist Party set itself to rouse public opinion against atmospheric nuclear testing over the Pacific ocean. My employer Frank Haigh was a regular speaker at protest meetings organised by the party in Auckland. Just as my junior role in the office required me to fetch Frank's lunch, so I was called on to help distribute the leaflets which denounced the testing and announced the times of the rallies and marches.

I did not enjoy this part of my job. I had no wish to be labelled a 'com'. I knew too that the limitation of the protests to atmospheric testing in the Pacific let the Russians off the hook. But I thought enough of the cause not to spurn support for it.

I used to puff my way up from the office of Haigh, Charters to the squalid headquarters of the Communist Party. I remember the *frisson* of apprehension I felt each time I entered. I'd heard enough tales of surveillance to believe I was being watched. Nor was there anything within to attract. Tucked away in a seedy arcade, the office offered no welcome. Neither did the staff, who sat among piles of boring-looking books, stacks of unsold party newspapers and always, in what seemed to me to be a denial of doctrine, the weekly racing tip sheet, *Best Bets*. Gathering up my stock of handbills, I left. I delivered the leaflets, I marched in the rallies and I went to the meetings.

The nuclear-free movement had an uphill struggle in the early 1960s. Although eighty thousand people signed a petition in 1963 calling for an end to nuclear testing in the Southern Hemisphere, there was nothing in New Zealand to match the mass protests organised by the Campaign for Nuclear Disarmament in Britain. The marches here were poorly attended; we counted it a successful meeting when the numbers reached the hundreds.

The United States had a test site at Christmas Island in the Pacific. It was bomb testing at Easter on Christmas Island that led to the public protest meeting I now best remember. The blasphemous combination sparked more than the usual interest. The smaller of Auckland's two town halls was almost filled with trade unionists, the women members of the League for Peace and Freedom, academics and a heavier-than-usual representation from the churches. The speakers all denounced the testing and were dutifully applauded. There wasn't a lot of passion in it — the audience had come convinced and needed no conversion.

The meetings never gave anyone much trouble. The police were not called on and the government was unmoved. Indeed, the gatherings probably gave comfort to the authorities because of the small support they attracted. The Auckland morning paper could always be relied on to alert its readers to the presence of communists among the speakers. While some were of Marxist leaning, most came from the churches or had a background of humanitarian concern. Almost all those protesters I met reeked of sincerity and goodness. They were indulged by the authorities and tolerated, even respected, by the wider public for their standing in the community and for their record of good works. The cause itself and their commitment to it left small impression. They hadn't learned to lobby or to cobble together any coalition of support.

The efforts of the Communist Party to make the nuclear-free movement its instrument were dismally unsuccessful. I was too often the witness of its insularity and incompetence to have any doubt of that. Its members lived as aliens in their own country, isolated and ineffective. The great irony is that the nuclear-free movement in New Zealand was finally set alight not by the efforts of the communists but by the actions of the Western powers.

American intervention in Vietnam created the first genuinely popular protest movement in New Zealand. The United States asked that its small but undeniably respectable ally make a show of support by taking part in the war. The National Government complied. The war served no obvious New Zealand interest; for the first time in its history the country was involved in a conflict it was not unpatriotic to oppose. A protest movement emerged whose national pride was obvious and whose sincerity was not easily questioned. It learned to test its powers of organisation and persuasion against the government. Any seditious or communist elements were rapidly overwhelmed by the mainstream. Protesters abandoned the limitations of musty halls and painfully duplicated circulars. They took to the streets, captured the churches, absorbed the universities. The papers were full of them.

The swelling tide of protest against New Zealand's involvement in the Vietnam War swept up the small nuclear-free movement, absorbed its energies, and for the time being obscured its aims. But all the energy, idealism and skill the protest movement had thrown up came very much to a loose end when the war in Vietnam petered out. The movement was left, as it were, looking for a cause to champion.

France in the meantime continued its programme of nuclear weapons testing in the sky above Moruroa Atoll in the South Pacific. This was not the immediate focus of any popular protest, although public concern about its harmful effect was widespread.

There was in fact no reason for protest. In 1972 the National Government was defeated; the new Labour Government decided to capture the mood of the country and become a protester itself by making a commitment to international action against nuclear testing. In a striking move designed to focus world attention on the testing at Moruroa, the government sent two of its small fleet of frigates into the test zone. A member of the Cabinet was aboard one of them in July 1973 as the French tested a five-kiloton weapon about twenty miles away, above the atoll. The gesture was reinforced when, together with Australia, New Zealand challenged France at the International Court of Justice at The Hague as to its right to explode nuclear weapons in the atmosphere. The case was adjourned when France gave way to international opinion and announced that it would in future carry out the tests below the surface of the atoll.

The Labour Government's next move was to set out to win international support for the creation of a nuclear-free zone in the South Pacific. Its proposal was adopted in principle at a meeting of the member countries of the South Pacific Forum in July 1975 and received the endorsement of the United Nations General Assembly later the same year. But that was as far as it went. In November 1975 it was time again for a general election. Labour lost, the National Party won, and no more was heard of the nuclear-free zone.

The Labour Government's concern for the impact of nuclear testing on the environment did not bring it into direct conflict with the United States. Conventionally powered vessels belonging to New Zealand's American ally paid occasional calls at New Zealand ports. Some of them may have been nuclear-armed, but the possibility aroused no public concern. The visits of nuclear-powered vessels would, in the climate of the time, have caused some consternation, but none came calling. Their absence was not, for most of the government's term in office, a cause of friction. It was not until 1975 that the United States formally accepted legal liability for damage caused by accidents involving its nuclear-powered vessels, and having done so, once again expected its ally to welcome their visits.

If the United States put pressure on the Labour Government

in its last few months in office to accept a resumption of visits by nuclear-powered ships, I did not hear of it. I was the Labour Party's candidate in the Hobson electorate and was not in the confidence of members of the government. Hobson was a large rural electorate, far in the north, and it was in the nature of New Zealand politics that the Labour candidate was not going to win it. For most of the voters I was living evidence of the creeping socialism the National Party warned about. Yet curiously enough, at the old settlement of Kerikeri there was one of the nurseries of the nuclear-free movement. Its warmth, its music and dramatic societies and its acceptance of the exile had made the town a magnet for the unorthodox. Many faiths prospered. Among Anglicans and Methodists were gathered Christian Scientists and Quakers and a smattering of old Christian socialists. Many among them had begun to question the alliance between New Zealand and the United States, with its underpinnings of nuclear ships and nuclear weapons. These pacifists of Kerikeri were very old and very idealistic. They were much given to prayer, invocation and argument: the theistic equivalent of the secular communists. They did not reflect the mainstream of New Zealand opinion about nuclear issues, and in that sense the Hobson electorate perfectly reflected the whole of New Zealand.

I was soundly beaten. I went back to Auckland and my legal practice.

Then there loomed on the horizon the prospect of sudden death by nuclear explosion or lingering death by leakage of nuclear radiation. The newly victorious National Government, led by red-baiting, round-headed Robert Muldoon, ever anxious to win the favour of the United States, invited the visit of the nuclear-powered cruiser USS *Truxtun*. Here was the spark that set alight the protest movement.

When it was first understood that the United States was anxious to resume its programme of visits by nuclear-powered ships, a protest group was formed in Auckland, calling itself the Peace Squadron. Its guiding spirit was a liberation theologian, the Reverend Dr George Armstrong. An engaging man of great vitality and considerable scholarship, he invited the churches and the non-religious peace groups to join him in mounting a marine protest. The Peace Squadron planned to use small boats to sail the channels of Auckland Harbour and physically bar the way of the intrusive nuclear technology. So began a series of water-borne demonstrations that were to mark the visit of every nuclear-powered vessel.

I knew George Armstrong and many of his fellow demonstrators through church activities, through my brother, who took to the water to try to defeat the intentions of the United States Navy, and through the potters and artists, friends of my brother, who had decided to commit themselves to protest against the visits of nuclear-powered vessels. Among them were many of those whose instinct for protest as a political activity had been honed in the demonstrations against the Vietnam War. I made no attempt myself to take part in the water-borne protests: although I'm fairly buoyant, I had no sailing experience. But it was understood that the demonstrators were going to find themselves in confrontation with the police and that some of them would need a legal defence. I warmed to the thought.

In October 1976, the nuclear-powered aircraft carrier USS *Long Beach* sailed towards Auckland. It came for what the Americans called rest and recreation. If its visit had any military purpose beyond the need to impress the public with the destructive powers of our American ally, I was not aware of it. Nonetheless I resented its visit; I simply didn't like the presence in the harbour of nuclear weapons. The Peace Squadron gathered together a flotilla of small craft and set out to sail across the narrow channel down which the *Long Beach* must enter the harbour.

On the morning of its arrival there was a flurry of activity. The harbour authority had designated certain areas off limits and dispatched its vessels, its officers and the police on board, to make sure the channel was cleared. Small boats darted across their path. In the shadow of Rangitoto Island at the entrance to the harbour, dwarfing its opponents and its guardians both, bulked the *Long Beach*. Into the fray set off the yacht *Dolphin*, captained by its owner Phil Amos, once a minister in the Labour Government. *Dolphin* sailed up the channel towards the advancing *Long Beach*, then Phil turned his yacht around and motored slowly down the channel back to Auckland, the vast vessel looming ever closer behind him. Not long after he made his turn the police intervened; the *Dolphin* was shepherded out of the channel and captain and crew were taken on board the police launch.

A police constable was put on board the *Dolphin*. 'Bloody hell!' the constable was heard to cry. 'Don't leave me here. I can't steer this!' This he proved by promptly going aground on a sand bank. The police launch turned around and Phil helped remove the yacht. Under official escort he sailed the *Dolphin* to the Admiralty Steps, a pier at the foot of Queen Street, right in the hub of the city, named for the landing place of the admiral's barge when it came across the harbour from the Devonport Naval Base. Here were gathered the ambassador of the United States, Armistead E. Selden Jr, his staff and local dignitaries, waiting to greet the captain of the *Long Beach*. With all the dignity he could muster, Phil embarked from the *Dolphin*. Armistead E. Selden Jr went up to him. He wrapped both arms around him. He embraced him, firm in the conviction that Phil was among the advance guard of the welcoming party. A police sergeant came up and pulled Phil away from the ambassador. 'This man is under arrest,' he said, and took Phil off to the Auckland Central police station, leaving behind a bewildered Armistead E. Selden Jr, who was himself somewhat of a curiosity, as he served three presidents in New Zealand, all of totally different persuasions.

Phil was charged with obstructing the police.

Charges of obstruction were the bread and butter of the Auckland Magistrates Court. Phil's case understandably had attracted a great deal of public attention. It seemed to me important to take the charge against him out of its usual context of minor offending and make it more of a political issue. There was called in evidence an official of the Ministry of Foreign Affairs, who told the court that the *Long Beach* had been invited to visit New Zealand by the government, so that being lawfully in the harbour, it was entitled to proceed unmolested about its business. I rose to cross-examine. Were there, I asked the official, any nuclear weapons on board the vessel? The official would not say if the vessel was or was not nuclear-armed. Then, I continued, did the New Zealand Government know if the vessel was nuclear-armed or not? He did not say no. He did not say yes. He did not even say that he was unable to answer the question. He simply pleaded that he be allowed not to try.

This was my first meeting with the doctrine of 'neither confirm nor deny'. The United States did not acknowledge the presence of nuclear weapons on its vessels. Nor was it willing to announce that any of its vessels were not armed with nuclear weapons. The USS *Long Beach*, which seemed to me to be the embodiment of the nuclear death machine, was coyly cloaked behind the doctrine. I wanted to press my cross-examination, eager to draw the official into contradiction, but the police objected; the point was not relevant. The magistrate upheld the objection.

When the hearing was over, Phil Amos was found guilty. I lodged an appeal. As it happened, I could not take the appeal to the Supreme Court as I became a member of Parliament and gave up my legal practice, but in due course the Supreme Court found that Phil had been wrongly convicted.

The public debate about the visit of the *Long Beach* was fuelled not so much by the ship's capacity to deal death and destruction through its weapons systems as by its means of propulsion, which was thought to be dangerous. The USS *Buchanan* visited Auckland three years after the *Long Beach*. The *Buchanan* was capable of carrying nuclear weapons and may or may not have been carrying them when it visited, but it was not nuclear-powered, and so its stay in Auckland passed without protest.

The visits of nuclear-powered vessels drew increasing numbers of demonstrators, and the protests gave me a sense of what might be possible. What in my experience more than ten years earlier had been a fringe activity was gaining ground to the point where it could easily become a major force.

The nuclear-free movement began to attract a growing number of women. Given the increasing reluctance of men to have anything to do with church activity, and the reliance of churches on women in their congregation, the nuclear-free groups in the churches were largely women's organisations. The nature of the movement altered; the hectoring, demanding rhetoric of earlier protest movements, the calculated affronts to political figures, the dialectical wording of resolutions from meetings, all gave way to expressions of concern and affection. The shouting of strident slogans was replaced by the presentation of gifts of flowers. Politicians were to be greeted or embraced and not to be railed at. The nuclear-free movement became what it should be, a movement of people who by their genuine commitment to gentleness and non-violence conveyed in the way in which they conducted themselves a glimpse of what a peaceful world might be like.

The protest activities made the visits into a political issue. Being for or against the visits was increasingly a measure of whether you were on the right or the left of New Zealand politics. Many of the opponents of nuclear weapons and nuclear propulsion came from the middle classes, and the nuclear-free movement was a path that took many away from the National Party which had been their political home. They turned to support of the Labour Party, now closely connected with the nuclear-free movement. But differences of opinion about nuclear issues were not the only reason for the National Party's loss of support among the middle-income groups. The unrelenting populism of the Prime Minister antagonised many. In 1976 Muldoon drove the Labour member of Parliament for Mangere out of office, hounding him with accusations of homosexuality. A significant number of these who had hitherto never thought of voting Labour found this behaviour insupportable.

I stood for Parliament as the Labour candidate in the by-election that followed the resignation of the member for Mangere, and was duly elected. Mangere was a safe Labour seat, close to the township of Otahuhu where I spent my youth. The nuclear-free movement held no interest for the voters of Mangere. My involvement in the Amos defence would undoubtedly have been noticed by the party faithful in terms of selection of the candidate but was blithely irrelevant in terms of the contest. The people of Mangere were poor, and the issues that concerned them were about the cost of living. The oppressive conduct of the Prime Minister in dealing with their former representative was a symbol of the difficulties which daily were heaped upon them. In electorates like Mangere, if you had no home (and many didn't), if you lived in a car and your daughter was pregnant, you didn't worry about the effect of pollution in the tidal marshes on the stilted heron. Nor could you be very concerned about the explosion of nuclear devices on Pacific atolls. I fought a campaign on the concerns of people who were struggling and, once elected to represent them, went off to Wellington to join my new colleagues in the parliamentary Labour Party.

2

Prodding the Sacred Cow

PARLIAMENT, WHEN I ENTERED IT, was dominated by the scourge of unpopular minorities and inviter of nuclear vessels to New Zealand, Robert Muldoon. Prime Minister for little more than a year, he had already and almost single-handedly changed the nature of New Zealand politics.

Muldoon became leader of the National Party in 1974, when the party was in Opposition to the Labour Government. Having found civility ineffective in the parliamentary contest, National chose Muldoon the brawler to take the battle to the government. He was, in one sense, an unlikely choice. The party of the farmers and the urban professionals was not the natural home for a self-made product of the working class. It made no difference; his street-fighting politics worked. Aggressively he led National into the general election of 1975, insult and innuendo flying all around him. A bruising campaign was followed by a landslide victory for the National Party and a triumph for Muldoon.

Having tasted blood, Muldoon was determined to keep it flowing. His party's huge majority became for him a licence to break out of the traditional restraints of parliamentary government. The camaraderie of political life rapidly withered and ideals of courtesy towards opponents faded. The Members' Bar in Bellamy's, the parliamentary catering establishment, was once the dinner-hour haunt of parliamentarians of both parties. Now it languished for want of customers. It was unsafe for parliamentarians to be seen there lest suspicion arise as to their true allegiance.

The new Prime Minister was an acute analyst of popular sentiment and by that unfailing light he conducted his government's foreign policy. The hapless subordinate who occupied the position of Minister of Foreign Affairs was usually left to pick up the pieces.

Muldoon sometimes made capital out of the warmth that many New Zealanders instinctively felt towards the countries of the old British Empire. He lent one of New Zealand's small fleet of elderly frigates to the British to help them fight the war in the Falklands. The rest of the time he went around picking quarrels. He sparred with the leaders of emerging nations in Africa, once telling them that they didn't even pretend to be logical. He told the United States that its defeat in Vietnam was the result of its unwillingness to use nuclear weapons. One American president he referred to as 'the peanut farmer'. Japan was a favourite target. The year I went into Parliament he was promising to drag Japan kicking and screaming into the twentieth century.

I received my baptism of fire in a Parliament that had left behind its traditional bipartisanship in foreign policy. I found myself in a culture which obliged the opposing parties to agree on nothing.

As the newest member of the Labour Party's parliamentary caucus, I joined the survivors of the 1975 election. Having briefly tasted office, the party had been humiliatingly defeated. We were led by Bill Rowling, principal target of the relentless baiting of the Prime Minister. There was little that was electrifying about Labour's caucus. It met weekly when Parliament was sitting to discuss party policy and parliamentary business. At the first meeting I attended there was passionate argument about soap and towels, and the need for locks on the doors of the shower block in the Opposition quarters. This debate carried on the following week. Then earnestly it was rejoined, and again after that. Most of the party's members of Parliament had spent far longer in Opposition than in government; almost by reflex action they opposed the government. In turn they were mauled and worried by Muldoon. We were a party beleaguered.

For all that, we were a party with a proud history. The New Zealand Labour Party was founded by trade unions to give their membership a voice in Parliament. Once a socialist party, it was unable to win a parliamentary majority while proposing, or threatening, the expropriation of private property. It gradually modified its policies until it offered a mildly left-wing prescription, doubtless an affront to the body politic in a country like the

United States but comfortably familiar in the rest of the English-speaking world. When I joined the parliamentary Labour Party there were still many trade unionists among its ranks, but there were also academics and schoolteachers and lawyers. At each general election (which by law had to be held every three years) the representatives of the professional classes grew in number.

In this, the parliamentary Labour Party reflected the party outside Parliament. Prime Minister Muldoon's relentless assaults on liberal sensitivities attracted towards the Labour Party the gentler souls among the public. There were those who wished for tolerance in race relations; there were the advocates of parliamentary reform, who hoped that reform might serve as a restraint on populism; there were opponents of visits by nuclear ships. In short, there were many who ten years earlier were not among those whom the Labour Party would have counted among its supporters.

Having chosen their new political home, Labour's recruits found themselves caught up in the ancient contest between the party outside Parliament and the party in the caucus. The party outside Parliament throbbed with idealism. Its ordinary members paid to take part in its meetings. There was a long history of distrust between those who saw themselves as sacrificial disciples of principle and those who, being elected to office, were paid from the public purse for their efforts. Whenever Labour formed the government the conflict heightened. Ordinary members were always on the lookout for party principle disregarded or dishonoured by members of Parliament in search of crude political advantage. When I joined the parliamentary caucus I found many of my colleagues deeply suspicious of the growing membership of the party outside Parliament. Their hold on the party was being wrested from them by the amateur enthusiasts.

Each year the Labour Party held a conference, attended by delegates from the party's branches, by the representatives of its associated trade unions, and by its members of Parliament. Each year the conference, in its ritualistic fashion, called the party's parliamentary wing to order. At most conferences, the votes of the parliamentary wing were outnumbered roughly eight to one by votes of other delegates. Parliamentarians who could not win an argument on its merits would lose it in a ballot. That wasn't to say that issues were finally settled on the floor of the conference. Under the party's wonderfully convoluted constitution, its policy was not made by its annual conference, but by a much smaller policy council. Members of Parliament made up nearly half of the council's membership, and often, to the fury of the ordinary members, translated the idealism of the conference into the baser coinage of everyday politics.

On issues of foreign policy, the annual conference was always good for a contest between the party in Parliament and the party outside it. Delegates to the conference regularly called for New Zealand's withdrawal from ANZUS, the military alliance between Australia, New Zealand and the United States. Some of the advocates of this measure were supporters of the nuclear-free movement, and were distrustful of an alliance which brought nuclear vessels to the country's harbours. Others saw the alliance as an unnecessary limit on New Zealand's independence. There was undoubtedly an element of hostility towards the United States, some of it ideological, some of it arising from the belief that New Zealand would not have taken part in the Vietnam War but for its ties to ANZUS. Whatever the motive, the delegates were to the alarm of the parliamentary caucus disturbing one of the great sacred cows of New Zealand politics.

The ANZUS treaty was signed in 1951, when New Zealand was seized by fear of Japanese recovery and uneasy at the prospect of expanding communism. Seeking the protection of a powerful guardian, it looked to have its security guaranteed by the United States. In this, New Zealand and Australia were of one mind. The United States, understandably, was cautious. It declined to commit itself to a treaty that would lead to inevitable American intervention in any regional conflict involving Australia and New Zealand. The three countries eventually signed an agreement to 'maintain and develop their individual and collective capacity to resist armed attack'. They agreed to do this through 'continuous and effective self-help and mutual aid'.

When it came to the crunch, the treaty partners would talk. Under Article III of the treaty, the partners bound themselves to 'consult together' whenever the 'territorial integrity, political independence or security' of any of them was threatened. Under Article IV, each treaty partner recorded its recognition that armed attack in the Pacific on any of them would be dangerous, and each declared that it 'would act to meet the common danger in accordance with its constitutional processes'.

From the time the treaty was signed, the United States was portrayed by conservative politicians in New Zealand as the country's inevitable defender, from whatever quarter it might be threatened. Many New Zealanders, remembering the Japanese invasion of the South Pacific, were genuinely grateful for this assurance. A majority of the public came to be firm in belief that safety depended on membership of ANZUS.

The new alliance enjoyed a good press. It had acquired its own mystique; it was a political totem, a symbol not only of New Zealand's military association with the United States but of all the political and economic ties that bound a small country to a superpower. It was drummed into the public by politicians that ANZUS was the key not only to New Zealand's security but to friendly relationships with the United States. ANZUS, it was argued, gave New Zealand the ear of the most powerful decision-makers in the world. Feeling about ANZUS ran deep. Opinion polls consistently showed clear majorities in favour of alliance membership, even after the issue was clouded by controversy about visits by nuclear-powered vessels. The result was that the last place on earth likely to be threatened by nuclear weapons was in the 1970s in active alliance with a great nuclear power.

That is not the same as saying that the popularity of the alliance depended on the proximity of the American nuclear arsenal. The doctrines of nuclear deterrence were actually some embarassment to the conservative politicians of the National Party who governed New Zealand. Nobody, of course, liked nuclear weapons. Nobody liked to be told that a country's security ultimately depended on them. Then there was the fact that New Zealand, for all the deep anxieties of some of its population, was not actually at risk of imminent invasion. The government's military advisers had invented the concept of 'forward defence' as a reason for sending soldiers thousands of miles away to South-East Asia. This was the closest they could get to identifying a military risk. All in all, the case for nuclear weapons in New Zealand was very hard to make. The National Party got around this difficulty by telling the public that ANZUS was not a nuclear alliance; it just happened from time to time that the country played host to American nuclear vessels.

As far as I was concerned, deterrence in the South Pacific was more than dangerous, it was absurd. Nobody could for a moment imagine that the United States would risk its people, and the world's, by defending its small and distant ally with nuclear weapons. No invader, if there ever was one, would be held back for a moment by such an unlikely possibility. The nuclear element in the alliance I found distasteful not only for what it represented but because its presence in New Zealand was a dismal pantomime.

If its nuclear component was abandoned, I saw no reason for New Zealand to withdraw from ANZUS. The weapons were the problem, not the treaty itself. Its wording in no way demanded their presence. As a plain political calculation, the Labour Party might as well go in for self-immolation as say goodbye to ANZUS. I argued against withdrawal from the alliance at party conferences and delegates hissed in ritual disapproval.

In the end, the call for withdrawal from ANZUS did not make its way into Labour's policy. Between conference floor and election platform, the plan was put aside. Although we fought the 1978 election on a promise to close the country's ports and airports to all nuclear-powered and nuclear-weapon-carrying craft, we assured the voting public that the ANZUS alliance would be unmolested by any future Labour Government. But here lurked uncertainty. As the last Labour Government found out, the United States insisted that its nuclear-powered vessels be free to call at New Zealand's ports. To ban them must at the least put a question mark over New Zealand's continuing participation in the alliance.

Prime Minister Muldoon was well aware of the political opportunities given him by American reservations about Labour's nuclear-free policy. Knowing the electorate's love for the ANZUS alliance, he set about prodding the sacred cow into motion in the hope it would trample his opponents. In 1976 he released a memorandum signed by the Secretary of Defence (the government's principal civilian adviser on military matters) and the Chief of Defence Staff. It was addressed to Bill Rowling when Rowling was Prime Minister. According to the memorandum, the government's advisers had been conferring with other officials in Australia and the United States and had carried home the hard word from Canberra and Washington. Exception had been taken there to the Labour Government's promotion of a nuclear-free zone in the South Pacific. It was, they said, 'unwise to test the ANZUS relationship in the manner in which we appear to be doing'. Muldoon rejoiced in this revelation, hailing it as proof that the nuclear-free policy put ANZUS at risk.

He was not alone in identifying the nuclear-free policy as a potential political weakness; some in the parliamentary Labour Party shared his opinion. The Labour caucus when I joined it was much divided over ANZUS. A few were opposed to New Zealand's membership, while others were intensely committed to it. Those latter would for preference discard the nuclear-free policy in favour of the alliance. I was among those who hoped we wouldn't have to make the choice.

At each general election and as new members of Parliament succeeded old, the nuclear-free sentiment grew stronger in the caucus. It went from strength to strength in the party outside Parliament. And its growth was greatly aided by the extraordinarily inept interventions of the American embassy.

In 1982 the party conference passed its usual resolution calling for the exclusion of nuclear-armed and nuclear-powered vessels from New Zealand ports. The American embassy issued a press release, claiming that 'our reading of the New Zealand electorate gives them a lot more credit than some people apparently do for their understanding of global issues and their sophistication about things nuclear'. Such statements I saw as a deliberate affront; the embassy must have known that its disapproval would have no influence at all on the party conference. It can only have spoken out in the hope of damaging Labour's standing in the eyes of the public. This I thought was a miscalculation; New Zealanders were no more fond than any other nationality of outside interference in their domestic disagreements.

I had to give more thought to the practical implications of the nuclear-free policy when, in 1979, I was elected deputy leader of the parliamentary Labour Party. Early in 1983 Bill Rowling stepped down from the leadership, and the parliamentary caucus chose me to replace him. How that happened is another story. In the meantime, the Labour Party, having lost the general elections of 1978 and 1981, was now well placed to win in 1984. Its organisation outside Parliament was stronger than ever, its parliamentary caucus was younger, it had more women members, it was more enthusiastic. Muldoon's street-fighting style, once so effective, began to pall on the electorate. The economy stagnated. The Labour Party would soon be the goverment.

As leader of the Labour Party I simply took it for granted that we would ban nuclear weapons from New Zealand as soon as Labour was elected. That was a deep-felt and instinctive understanding. I had not changed my mind about the threat posed to the planet by nuclear weapons since the evening I saw the nuclear aurora fill the sky above my home. It seemed to me wrong for a country like New Zealand to play host to nuclear missiles, even if they came on ships full of recreationally minded sailors. To welcome them into our ports was to enter into compliance with military strategies which promised uneasy peace at best and utter destruction at worst. There might be little that a small country could do to influence the thinking of the nuclear powers, but New Zealand did not have to surrender to their practices. We could, if we wished, stand aside from the endless escalation of nuclear armaments.

In this I was at one with a large majority of my parliamentary colleagues and most of the active membership of the Labour Party outside Parliament. Nor did I see the banning of nuclear weapons, set aside from any question of the impact of their exclusion on the ANZUS alliance, as a political liability. Some people in fact were going to vote Labour only because of the party's nuclear-free policy. Others would support the nuclear-free policy because they were going to vote Labour anyway and that was part of the package. Banning nuclear weapons would be a simple step to take; in its way, an assertion of independence and an example to others. I was sure that the promise of it would in

Opposition be popular and in government be implemented with the support of the public.

I then had to think about how Labour might in government carry out the nuclear-free policy while staying in the good offices of the United States. During the 1981 election the Labour Party, proclaiming once again its opposition to the visits of nuclear vessels, asserted in its statement of policy that its nuclear-free stand was not contrary to the ANZUS treaty. However true in law that claim might be, it did not stand up well against the obvious displeasure of the United States. If New Zealand was to be both nuclear-free and stay in ANZUS, there was only one course open to us. We had to persuade the United States that it would be to its advantage to accommodate a nuclear-free New Zealand in the ANZUS alliance.

The immediate difficulty in bringing the Americans round to our way of thinking was that the nuclear-free policy seemed squarely aimed at the United States and the United States only. From time to time, ships of the Royal Navy which may well have been nuclear-armed called at New Zealand ports. These, too, would have been banned in terms of Labour's policy, but their infrequent visits attracted not a fraction of the uproar that greeted American nuclear vessels. No Soviet naval vessels called. To counter suggestions of anti-Americanism, I endlessly rehearsed New Zealand's friendship for the United States whenever I talked about the nuclear-free policy. I stressed our many common interests — the political equivalent of abhorring the sin while loving the sinner. Sinners do not like to be reminded they are sinning, but it was the best I could manage.

Another difficulty in reaching an accommodation with the Americans came from what seemed to me to be a flaw in the reasoning behind Labour's nuclear-free policy. The party promised the country that it would ban what our policy called 'nuclear warships or aircraft'. By this was meant any aircraft armed with, or carrying, nuclear weapons, and any naval vessel armed with, or carrying, nuclear weapons or powered by a nuclear reactor. The ban on nuclear weapons I welcomed. The ban on nuclear propulsion gave me problems. Only the United States brought nuclear-powered vessels to New Zealand. I thought it sensible that the party's policy should take account of the danger to public safety posed by nuclear reactors, and that there might well be grounds for banning nuclear-powered vessels from our ports on that account. But it seemed idle to raise that kind of concern into a pillar of foreign policy. We should be able to distinguish between the need to guard against the immediate dangers of nuclear reactors, and our wish to counter the threat posed by political decisions to build, deploy and threaten the use of nuclear weapons. A stand against the arms race was the legitimate concern of foreign policy. But I could not see how the arguments I mustered against the deployment of nuclear weapons could properly be used to justify a ban on nuclear propulsion. Shutting out their nuclear-powered ships for the same reasons we shut out their nuclear weapons seemed to be offering an unnecessary affront to the Americans. If we continued to lock propulsion and weapons together and did not distinguish them, I was not sure how I could persuade the United States of the essential rationality of our policy.

I was leader of the Labour Party for only a few weeks when I announced that I wanted the party to review its policy on nuclear propulsion. I wanted the policy to allow for visits by nuclear-powered vessels if acceptable standards of safety were met.

There followed a hard lesson in practical politics in which I scarcely opened my mouth before the argument was lost. I did not, when I proposed this rewriting of policy, have command of the Labour Party outside Parliament. I had served no long and arduous apprenticeship in the party's lower ranks. The party's parliamentary leader was elected by the parliamentary caucus, and it was caucus endorsement, not grassroots enthusiasm, which thrust me to the front. Many ordinary party members indeed resented my replacement of the well-regarded Bill Rowling. My desire to alter the nuclear-free policy was met with deep suspicion; I had not prepared the ground. It was a misjudgment. Press, radio and television fed happily on stories of party differences.

If there is one quality more than any other that is marked down heavily in New Zealand politics it is the appearance of disunity. It is better to be clinging to the juggernaut as it heads for the cliff than to be seen to be arguing about the direction you are going. The Labour Party, with its native attraction for idealists of all persuasions, suffers perpetually from its inability to keep its disagreements out of the public eye. The argument about nuclear propulsion, flaring immediately into the open, was immediately damaging. I saw no point in persisting with it. If Labour threw the 1984 election because we were fighting among ourselves, we wouldn't have the chance to put any policy into practice, however flawed I might think it. So I bowed to the inevitable and accepted that there were to be no qualifications on our nuclear-free policy.

Labour was now treading a difficult path. The looming prospect of government was hedged around by American displeasure at the nuclear-free policy, with all the doubt that created about American willingness to come to New Zealand's military assistance. Labour could not fight the 1984 election acknowledging that its nuclear-free policy put ANZUS into question. I knew exactly what would happen. Senior members of the National Party would glimpse the dark and threatening shapes of Soviet submarines beneath the surface of New Zealand's coastal waters. Blurred photographs taken by military surveillance aircraft would warn the public of the constant watching presence of the enemy. In true Cold War fashion, members of the Communist Party of New Zealand would be identified and their plotting uncovered. A few months of that and we might well have panic. Labour needed something more than a declaration that the nuclear-free policy was not incompatible with alliance membership. The National Party and the American embassy were only too happy to point out to the voters that a nuclear-free New Zealand could not be in ANZUS.

To add to our difficulties there was now a head of steam building up in the Labour Party outside Parliament for withdrawal from ANZUS. This time the surge of opinion seemed unlikely to be appeased by a ritual call for withdrawal at the annual conference. Having put to rest the argument about nuclear propulsion, it seemed as if we might fling ourselves into a far more dangerous public contest about the future of ANZUS.

Bill Rowling was the Labour Party's spokesman on foreign policy. Privately, he summed up the party's options. We could assert our continuing support for the ANZUS alliance. This option, unlikely to win favour at the annual conference, was increasingly suspect given the discrepancy between the nuclear-free policy and the American view of how the alliance ought to work. Our next option, to withdraw from the alliance, was bound to be popular at the conference but would lead, as I saw it, to electoral disaster. Rowling then offered a third possibility. The Labour Party should fight the election on an undertaking to seek a review of the ANZUS alliance. The next Labour Government should try to reach agreement with Australia and the United States to broaden the scope of the ANZUS treaty. A new treaty would provide for co-operation between the partners on economic, cultural and political issues, reducing the importance of the military side of the relationship.

Rowling's third option might persuade members of the party who favoured New Zealand's withdrawal from ANZUS that an alliance of a different character could be worth staying with. Easily accepted by members who recognised the electoral appeal of ANZUS, it was basically a holding position. Rowling wrote in a memorandum to the party's policy council that 'the review option, it can be argued, has the advantage of leaving open the possibility that Labour will be able to obtain its nuclear weapon free zone objects and remain in ANZUS'.

The proposal seemed essentially reasonable. It foresaw an association between New Zealand and the United States which did not depend for its existence on the visits of nuclear ships, and it gave the alliance partners the opportunity to decide how they might best jointly promote their common interests.

I wasn't sure if Bill Rowling would be able to persuade the annual conference to curb its enthusiasm for withdrawal from ANZUS. In the end, however, he succeeded brilliantly. He called in every credit he had with the conference delegates. In their affection for him they listened to his argument and voted to support his proposal that New Zealand seek a review of ANZUS.

So it happened that the Labour Party fought the 1984 election pledged to make New Zealand nuclear-free and pledged, as our policy put it, to 'renegotiate the terms of our association with Australia and the United States'. To make sure that no politician took the opportunity of the renegotiation to barter away the party's principles, our policy stated, in the belt-and-braces language beloved of party activists, that the renegotiation must encompass 'New Zealand's unconditional anti-nuclear stance' and 'an absolute guarantee of the complete integrity of New Zealand's sovereignty'. I was happy enough. Our policy coupled the exclusion of nuclear weapons with an approach to ANZUS that would not unduly alarm the public and formed an honest basis for an accommodation with the United States.

That we would soon be looking for an accommodation I was now certain. At the start of 1984, the National Party was in a state of confusion. Muldoon, its leader, thrashed about him. His subordinates plotted against him. Opinion polling signalled that Labour was heading towards government. I would be Prime Minister, and in New Zealand's small government, where every Cabinet minister had more than one responsibility, I was going to appoint myself Minister of Foreign Affairs as well. I knew there would be some kind of testing between New Zealand and the Americans, but what the outcome would be, I had no way of knowing.

3

In the Lion's Den

I WANTED TO VISIT the United States in 1967, when my law studies were finished and I set out to see the world. My application for a visa was refused. The consul's office in Auckland gave no reason. I suppose my involvement in protest against the war in Vietnam and in the legal defence of some prominent demonstrators had come to the embassy's attention. Instead I went to London. When I arrived I was taken into custody by officers of the Special Branch. My luggage was searched and questions asked at length about my travel plans. I rose in my own estimation at the thought that my modest efforts at political protest had reached the notice of the American and British authorities, but looking back on it now, I think that Special Branch might have mistaken me for somebody else.

My impulse towards protest came instinctively from a belief that people should not be the oppressors of others. I objected to foreign powers testing nuclear weapons in the South Pacific against the wishes of the inhabitants. It was an affront to me that New Zealand's soldiers were sent to Vietnam, as it were at the beck and call of a great ally. This seemed at odds with the democratic values in whose name the enterprise in Vietnam was launched. But the teachings of the far left held no attraction for me. I was tolerant of the Communist Party of New Zealand because I saw very clearly its essential incompetence; no prospect offered of the party's putting its dogmas into practice. (The party later proved beyond doubt its unprofitable adherence to unlovely doctrine when, detecting departures from the true faith in the governments of Russia and then China, it became the only communist party in the world to align itself with the regime in Albania.)

I knew that there was nothing pleasant about what happened if you were caught up in the thralls of Marxism as practised in

Eastern Europe. I was once in the Magistrates Court at Auckland when a man appeared charged with careless driving causing death. The maximum penalty for the offence was three months' imprisonment or a fine of four hundred dollars. The prosecutor read out the charge. The man in the dock was obviously distressed. Suddenly he shouted, 'Shoot me now! Shoot me now!' The prisoner was sincere in his belief that he was doomed. He was an East German; his mother was Hungarian and she brought him up to speak the language. After the uprising in Hungary in 1956 he somehow escaped to Italy, where he passed himself off as a Hungarian refugee. He came to New Zealand, where he found a job in a timber mill. Certain that attracting notice would lead to his uncovering, he lived a life of total seclusion. He worked at night; he occasionally went to the cinema. For the rest, he spent his time in his room or walked, usually in the evening after dark. This life he had lived for twenty years, during which the burden of oppression never fell from him. He was eventually convicted of the charge, fined two hundred dollars and disqualified from driving.

My unfavourable view of Marxist government was enhanced on the only visit I made to the Soviet bloc. In 1980 I went, as a member of Parliament, to an International Parliamentary Union conference in Czechoslovakia. Of the business of the conference I have little memory, but the exasperation of a journey behind the Iron Curtain was drearily unforgettable.

I travelled with Leo Schultz, a National Party MP who represented a deep-dyed, conservative, country electorate. We arrived at our Prague hotel. My room was dark and smelled of damp. The lavatory was down the corridor. The first time I used it the cistern collapsed and I was baptised in the Prague water supply. Leo, meanwhile, was living in luxury in the penthouse suite. I asked him how he managed it. He smiled. 'Whenever I'm in a communist country I send out postcards saying, "Freedom abounding, prosperity rampant, you should see this place, love, Leo."' He sent the cards to addresses he invented in New Zealand.

At the conference we ate caviar with dessertspoons under the eye of security guards. The conference over, the state travel agency refused to confirm my airline reservations. It cost me a bribe of sixty US dollars to buy myself a train ticket. At the border the train stood for hours while security guards poked and prodded and dogs snuffled through the carriages. When we finally arrived in West Germany, a working telephone came as liberation.

In my opinion of the totalitarian regimes of Eastern Europe, I don't suppose I was at odds with anyone I ever talked to at the Pentagon or the State Department. But I didn't end up at the same conclusions. To confront totalitarianism with more of the same did not seem right. It worried me that the United States put pressure on a little country like New Zealand to bring its ally into line with its wishes. Only a few years earlier America had expressed its displeasure at the Labour Government's proposal for a nuclear-free zone in the South Pacific. Such a zone might call into question American capacity to sail its naval vessels freely up and down or drop anchor in Pacific ports. The National Government, knowing America's wishes, had buried the zone. That to me was a denial of the right of the people of the South Pacific to decide their own destiny.

The National Party bowed to American wishes because it valued the presence of American naval vessels more than it valued the absence of the nuclear deterrent. What lay behind that judgment, I am not the one to tell. When called upon in public to justify its actions, National gave only one reason. New Zealand could not be left defenceless against 'the Soviet threat'.

'The Soviet threat' (the same that caused Russian submarines to appear in sight of our coastline when elections were in the offing) gave urgency to American arguments for New Zealand's submission. We heard how the Russians were expanding their activity on their North Pacific coast. Then they were busy at Cam Ranh Bay in Vietnam. New vessels were being built for their Pacific fleet. Russian delegations were reported to be tempting the governments of tiny South Pacific island states with offers of aid, laced with demands for port facilities and landing rights for Aeroflot. In face of this onslaught, New Zealand must surely see reason and abandon its possibly well-meaning, but certainly obstructive, efforts to keep nuclear weapons out of its backyard.

I don't know how seriously the National Party took this line of argument, but it was always what was trotted out in public. Others found it hard to swallow. I didn't consider we were in danger. Cam Ranh Bay was closer to Paris than it was to Wellington. The war in Vietnam was over, but the dominoes of South-East Asia had not, as was predicted, fallen. The island states of the South Pacific were unlikely to subject themselves to Marxist teachings while the Christianity of the missionaries still flourished. If the Russians had any purely material attractions to offer, they could be countered by economic assistance from Western powers.

Whatever the Russians might be up to in the South Pacific, and it wasn't very much, the presence in South Pacific harbours of American nuclear vessels wasn't going to make the least difference. New Zealand itself, a thousand miles distant from its nearest neighbour, was not going to become a focus of Soviet strategic planning. In my view of what was actually happening in the South Pacific, a threat existed only in the minds of those who saw any local difficulty or disturbance, wherever its origin, as part of the grand design of the Soviet Union.

I found it disturbing that American officials I spoke to seemed unable to talk about events in the South Pacific without framing them in terms of a global struggle between the United States and the Soviet Union. It puzzled me, too, because this perspective, and the sometimes overbearing and demanding behaviour that accompanied it, were at odds with what I'd heard about the United States itself. The picture I had of it was of a country essentially democratic, seething with concern for the rights of individuals, seeming indeed to live by the code of freedom it espoused. Its insistence that its allies play host to its nuclear vessels, however unnecessary those visits might be, struck a jarring note. Yet almost every American official I met told me that the visits must take place or the anger of the United States would descend on our heads.

In my early days in Parliament I met few enough of those officials. The American embassy in Wellington was noted for its cultivation of conservative politicians. Its hospitality flowed generously over them. It did not concern itself with politicians it placed on the left of the spectrum. I only once dined at the ambassador's residence. After the soup course came the news that an aircraft of Air New Zealand had crashed while on a sightseeing flight above the Antarctic, and the dinner was abandoned. Most of the embassy's dealings with the Labour Party's members of Parliament were conducted by lesser beings than the ambassador, labour attachés and the like. Nothing in their manner suggested that they had come to grips with a party whose aim had once been the public ownership of the means of production, distribution and exchange. They gazed upon the Labour Party as I might watch a Hindu festival, fascinated by the sound and colour, made uneasy by its alien character and utterly uncomprehending of it.

From time to time a senior official of the United States administration, or a senior officer of its armed forces, would visit New Zealand for talks with the government. In the normal course of events, these visitors would call on members of the parliamentary Opposition. One such visitor was Admiral William Crowe, who was then Commander in Chief, Pacific.

Bill Rowling was Leader of the Opposition, and in his office to meet the admiral there gathered Bill, me (in my capacity as Deputy Leader), Arthur Faulkner, who was Minister of Defence in the last Labour Government, and Mick Connelly. Mick was the leading light of the conservative faction in the caucus, a staunch advocate of every conservative cause. I once went campaigning with Mick in a little country town. An election was close and there was much more interest than usual in the meeting we were holding. I was supposed to be the star turn, and Mick was meant to say a few words of introduction. But the subject was the police, one of his favourite topics. He went on and on. Under a Labour Government the police would get more pay, more respect, longer holidays. It just didn't stop. It was mind-numbing. After forty minutes he had finished so I got up and said, 'We're gonna take a break now for ten minutes so you can all go out and join the police.' Mick brought the same strength of conviction to his defence of the ANZUS alliance. He knew well that it was only ANZUS that stood between New Zealand and 'the Soviet threat'. If nuclear ships in our harbours were the price of the alliance, then so be it. He must have been looking forward to his meeting with Admiral Crowe.

The meeting did not go well. The admiral began with a powerful dissertation, marred only slightly by his confusing New Zealand with Australia. He told us about new Soviet activity at Cam Ranh Bay, emphasising its closeness to New Zealand. His thesis was simple. Global security was indivisible. New Zealand could not remove itself from the world-wide struggle against communism. We must continue to play our part in the battle and shoulder our share of the burden. From there it was all downhill. Perhaps I looked bored. Bill Rowling had a quizzical way of speaking which seemed to irritate our guest. Arthur Faulkner was known to the admiral as a Minister of Defence who had refused to allow his country to be defended by nuclear-powered vessels. The admiral's annoyance at his failure to make any headway was becoming obvious.

Wanting to help, Mick Connelly earnestly enquired, 'Tell us who the real enemy is.' Utterly mistaking Mick's intentions, Crowe turned on him. 'Cut out the bullshit!' he snapped. Mick had the look of a whipped dog.

The State Department, meanwhile, was anxious to bridge the gulf of cultural misunderstanding. It regularly invited those it held to be opinion-makers to visit the United States to examine aspects of that country's life and culture at the department's expense. In 1982, still Deputy Leader of the Opposition, I accepted an offer to spend several weeks in America. I revelled in the extraordinary irony of an invitation to a country that had refused me entry fifteen years earlier, and I was eager to see at last the real world of the news bulletin. As was their practice, the State Department asked what I wished to accomplish in the United States. They would then arrange an appropriate programme. I told them that I wanted an insight into what had motivated a country to elect in one term President Jimmy Carter and at the next election President Ronald Reagan.

The search for the answer to this question took me to two towns that faced each other across a river in Illinois. They were known by their joint names of Sterling-Rock Falls. Here I stayed with a pleasant family, the father a physician and a collector of rebuilt classic cars. I went to a steel works and its manager showed me around. This man had not missed attendance at his local church since he was a boy, and he did not leave the town at the weekend so that he would never miss it. I marvelled at his devotion but wondered why the owners of the mill had entrusted their hugely sophisticated business to one of such simplicity. He showed me a new extension to the rolling mill. By New Zealand standards, it was a huge project. I asked my guide what sort of finance was available for it. None really, he told me. So how was it funded? To my surprise he told me that the extension had been paid for from retained earnings. 'But surely,' I said, 'the shareholders objected to the company's keeping back such a share of profits.' 'That's not a problem,' he replied. 'I own sixty per cent of the stock.' In Sterling-Rock Falls, Illinois, were nurtured the strong, simple feelings to which Carter and Reagan variously appealed: love of God, love of family, love of country, pride in thrift, pride in achievement, pride in citizenship. Good and evil here were easily identifiable. Carter had proved himself a disappointing guardian of these totems. Reagan spoke better for them.

The State Department sent me on a tour of official Washington. This pilgrimage I repeated at greater length in 1984. By then I was leader of the Labour Party in Parliament, and travelling at the expense of the New Zealand and not the American taxpayer. I stopped for a few days in the United States, at a time when the American bureaucracy must have calculated that there was at least a strong possibility of a change of government in New Zealand, and that the United States would have to deal with the new government's nuclear-free policy.

Among those I met in Washington was Vice-President Bush. Our meeting had an air of informality. The Vice-President, giving no impression that business was pressing on him, showed little inclination to keep to the timetable. Our principal topic of conversation was butter. It happened that the United States was planning, as part of an aid programme, to unload a substantial quantity of surplus butter in the Caribbean. This would have destroyed a trade New Zealand had built up over many years. Such a reminder of the extraordinary capacity of the United

States to damage New Zealand's economic interests, even while acting with the best of motives, I found deeply worrying.

We skirted around the Labour Party's nuclear-free policy. The Vice-President talked about Australia. There, a Labour Government had been elected, pledged, like the New Zealand Labour Party, to remain in ANZUS, but pledged also to assert greater independence in foreign policy and committed to adopt active measures of arms control. Australia was deeper into the alliance than New Zealand. As well as playing host to the visits of American nuclear ships, it had American installations on its territory, all of them the subject of much suspicion among the nuclear-free movement. The left wing of the Australian Labour Party found ships and bases alike distasteful. But when the new government got its bearings, it decided that it liked ANZUS just the way it was. Bush now observed to me that despite what he called the 'initial talk' in Australia, relations between the United States and Bob Hawke's government had been 'damned good'. Thinking it politic to avoid the obvious implications of this remark, I told him that people in New Zealand were just as well disposed towards the United States as were people in Australia. He might take from it what he pleased.

Bush I found pleasantly amiable. Many months later, when I was Prime Minister and the United States had given New Zealand the diplomatic cold shoulder, I met the Vice-President at a United Nations reception. I was much surprised, given the freezing of relationships between the New Zealand Government and official Washington, that he greeted me warmly, and I said as much to him.

'Oh no,' he said, 'it's not my job not to speak to you.'

Less engaging was my visit to the Pentagon. In this blank-faced building, little golf carts puttered along endless corridors among a warren of offices. I was taken to look at the ANZUS corridor. Here, among the flags of Australia, New Zealand and the United States were placed images of the various military operations in which the three countries had taken part. A diorama showed scenes of battle. The difficulty with all of this was that the conflicts depicted had nothing to do with ANZUS. New Zealand soldiers did not fight in Korea or Vietnam because the provisions of the treaty had formally been invoked. The integrity of no alliance partner had been threatened. The ANZUS corridor, it seemed to me, was a model of false pretences.

At the Pentagon and the State Department, officials spoke about the difficulties the Labour Party's nuclear-free policy presented to the United States. They chose their words carefully. They did not say that the ANZUS alliance would come to an end. They did not even say that a ban by New Zealand on American nuclear vessels would make alliance operations impossible. They said, quite matter-of-factly, that a ban would make alliance operations difficult. There was no undertone of menace, and no words spoken which even the most sensitive could read as a threat. Like the Vice-President, the officials took much comfort from the Australian experience. Here was a government that had flirted with notions of rewriting the alliance but had then seen the light. I recited to them the New Zealand Labour Party's position. If they wanted to recast it in the warm light of the Australian experience, that was up to them.

I couldn't see New Zealand going down the Australian path. The Australian Labour Party in embracing ANZUS was making a positive virtue out of nuclear deterrence. Australia's Foreign Minister, Bill Hayden, acknowledged that American installations in Australia were nuclear targets, but argued that the risk was worth it because the installations formed part of America's nuclear deterrent. (He also liked to argue that Australia could use a threat to close the bases as leverage against the Americans, a position which somewhat undermined their supposed value as a deterrent.) In other words, if there was an arms race, Australia was a starter.

Although visiting United States admirals might mix the two up, New Zealand and Australia were very different countries. New Zealand was much smaller and far less likely to identify itself with the big boys on the block. It's hard in New Zealand to believe yourself to be a critical part of anyone else's global strategy. But a lot of Australians are genuinely certain that the world would miss them if they weren't there. Many Australians look instinctively to the United States as friend and protector. If you ask New Zealanders which country apart from their own is most important to them, they won't usually cite the United States by way of answer, although most Australians, asked the same question, say just that. In New Zealand, the answer to the question is, most often, Australia.

The politicians elected to office in New Zealand and Australia duly reflected these contrasting perspectives. I've met many Americans I liked, but I'm not sure I ever met one I felt I really understood. I always look forward to going to the United States, but I think it's because I always expect to be astonished by what I find there. I certainly never dreamt of trying to make political capital out of any American connection I might be able to conjure up for myself. America is too big ever to be wholeheartedly embraced by the New Zealand public. The kind of love affair Australian public opinion had with the United States just wasn't possible here.

Bob Hawke, on the other hand, was just the man to encourage his country's infatuation with America. He wanted to mix with the big boys of the world, to be accepted as one of them. He fell easily into complicity with them, enjoying their patronage, their power, their influence. And in the world of the big boys, nuclear weapons were the ultimate aphrodisiac.

Despite my encounters in Washington, I remained unseduced. But I'd heard enough to be worried about what might happen if we managed to get ourselves into a squabble with the Americans. The spoiler was the ANZUS alliance. Since it signed the treaty, New Zealand had little history of freedom of action in its foreign policy. In the United Nations it voted with the United States more consistently than the Warsaw Pact satellites supported the Soviet Union. Attempts at independent action, such as the proposed creation of a nuclear-free zone in the South Pacific, had petered out. As far as I could tell, the United States was taking New Zealand's continuing acquiescence very much for granted, and wasn't likely to be happy if we ever questioned it.

The whole thing had to be turned around. Most New Zealanders wanted to be in ANZUS and lots of them wanted to be nuclear-free. But unless the United States changed its mind about the nature of the alliance, we weren't going to have it both ways. I had to rely on the review of ANZUS, which was a key plank in the Labour Party's policy. If New Zealand could somehow persuade the United States to focus on the real perils facing the South Pacific, then it might come to accept that nuclear weapons were quite inappropriate to the defence of the region. By trying to strengthen economic and cultural ties, I hoped to put the defence arrangement into softer focus and convince the United States of the potential value of low-level military co-operation among the three ANZUS treaty partners in the South Pacific.

I had some grounds for believing that whatever happened, the United States would not straight away tear up the ANZUS treaty and rule out any possibility of its renegotiation. As a practical matter, it was clear that America's occasional deployments of nuclear weapons in New Zealand served no military purpose. The weapons were sometimes brought on vessels that came to show the flag or visited for recreational purposes. Sometimes vessels called which had been exercising with New Zealand ships, but the tactics of nuclear warfare were not part of the exercises. No missiles were deployed in New Zealand and none that we knew of were pointing towards us. On the great nuclear chessboard, our little country didn't rank as much as a pawn.

Then, too, those whose opinion I respected had it that the United States was unlikely to deal too harshly with New Zealand. Bill Rowling, who had what I had not, experience in government and diplomacy, was sure that our ban on visits by nuclear ships would not mean an end to the ANZUS alliance. He wrote that the United States would not want to be seen bullying a small ally on the sensitive nuclear issue.

Finally, there was a practical restraint on the United States. If it tore up the treaty, it ended its alliance with Australia, which it obviously valued. It was unlikely to want to draw up a new defence treaty with Australia when the old one suited its purposes so admirably.

One way or another, the treaty was going to be around for some time. I had to find a way of changing it from a millstone around the neck of nuclear-free New Zealand. Nothing I'd learnt in official Washington, or heard from the American embassy in Wellington, made me think it would be easy. But I did not think it was impossible. And it didn't for a moment occur to me that we would have to reverse our policy. The people of Sterling-Rock Falls, Illinois, and the other places I'd been to in the United States convinced me that what I believed about America was true. The American commitment to democracy and all that went with it was far-reaching. It was unthinkable that the United States would demand that the New Zealand Government ignore the wishes of the people who elected it and change its policy.

4

Immovable Object, Irresistible Force

IT WAS A TELEVISION PERFORMANCE by Prime Minister Muldoon, who on the night seemed much the worse for wear, which gave the Labour Party the break it needed to make New Zealand nuclear-free.

It was not a good year for Muldoon. In the middle of the year his government ran out of money. There was no way that they could present a budget without disclosing that all that could be borrowed had been borrowed. The books could not be made to balance.

It seemed to many of us in the Labour Party that Muldoon was casting around for an issue which would justify a general election earlier than that expected in November, and thereby avoid the need to produce any budget at all. He must have decided that Labour's nuclear-free policy gave him a fighting chance. An unlikely representative of what had come to be called the peace movement put Labour's policy on the parliamentary agenda. The emotional spectrum of Richard Prebble, MP, ranged from the truculent to the incandescent. Though no pacifist, he was long active in the nuclear-free movement. In his electorate lay Auckland's Waitemata Harbour and the outlying islands of the Hauraki Gulf. Prebble was a prominent speaker at public meetings held in Auckland to protest against the presence in the harbour of nuclear vessels. He was also a useful parliamentary strategist.

Prebble was an enthusiast for introducing, and thereafter debating to the limits of the time allowed, private members' bills. By this device, Opposition members can seize parliamentary time to debate topics of their own and not the government's choosing. This infuriated the government and cheered up the Opposition, but in the normal course of events the bills did not pass into law. Using its majority, the government killed them.

In 1984 Prebble drafted a private member's bill which he called the Nuclear Free New Zealand Bill. If the bill ever became law, it would have three results. It would ban the entry of nuclear-powered ships and nuclear weapons to New Zealand. It would forbid the building of nuclear reactors (already in 1984 an unlikely prospect, but one which nonetheless caused agitation among many of the public). And it would outlaw the dumping of nuclear waste. The bill, when it came there to be discussed, proved a measure of the shift in the centre of gravity of the Labour caucus. No warnings of the affront it offered to the United States were sounded. No argument was made that the bill endangered the ANZUS alliance. No demand was made for a committee to be formed to study it. Mick Connelly glowered. Nuclear-free New Zealand had become an article of faith, and the Labour caucus dealt with articles of faith in one of two ways. It might subject a proposal to endless analysis, searching for the refinement that would show further the purity of the party's convictions. Or it might simply bow down before it, acknowledging a proposition seen as absolutely indicative of the party's position. Caucus duly bowed before the Nuclear Free New Zealand Bill.

There was, it must be said, an element of political calculation in the bill. The National Party had a working majority of one vote in the ninety-two-member Parliament. Two members of the Labour Party, disaffected by the party's decision not to adopt them as its candidates in 1984, had left the Labour caucus and now sat in Parliament as independents. If it rallied these two to its cause, the government's majority rose to three, a margin still narrow enough. It was the Labour Party's constant hope to splinter the National Party and lose the government its majority by tempting some of the more independently minded government members to vote with the Opposition on matters of principle. Given Muldoon's disregard for constitutional niceties, there were plenty of matters of principle around to vote on, but fierce party discipline had so far held the government's troops in check.

Two National Party members of Parliament were known to be

supporters of the nuclear-free movement. If the Nuclear Free New Zealand Bill attracted their votes, and the independent members voted with the Labour Party or (more likely) missed the vote, the government might suffer what in New Zealand terms would be the unthinkable embarrassment of a parliamentary defeat.

The day chosen for the parliamentary debut of the Nuclear Free New Zealand Bill was 12 June 1984. On the morning of that day Bill Rowling and I took the stage to present to press and public the Labour Party's foreign policy. Yes, I declared, making my announcement, Labour would renegotiate the ANZUS treaty to accommodate the party's nuclear-free policy if it won that year's election. Yes, said Bill Rowling, answering questions afterwards, the renegotiation would be pursued. He thought it highly unlikely that there would be a split among the ANZUS treaty partners or that Australia and the United States would end up in an alliance by themselves. Not so, said the Prime Minister, later that evening. The Commander in Chief, Pacific, and the American Secretary of Defense had told him that it would be difficult for the United States to fulfil its military obligations to New Zealand if its navy did not have access to New Zealand ports.

Later that day the parliamentary curtain rose on the Nuclear Free New Zealand Bill. Labour members pointed to the inability of the United States and the Soviet Union to agree to arms reductions, and the failure of multilateral efforts at arms control. It was all the more important for countries like New Zealand to take unilateral action against the spread of nuclear weapons. National members argued that the nuclear-free policy meant the end of ANZUS. Perhaps in an attempt to heighten the air of crisis, the National Party's Minister of Defence made the extraordinary statement that the government would not seek the Governor-General's consent for the bill if it was passed by Parliament (a formality needed before the bill became law). This remarkable proposal, a kind of suspension of the constitution, was little noticed among the constitutional outrages which in those days were the common currency of the Muldoon government.

The bill might have proved a parliamentary triumph for the Labour Party. Two of the National Party's members duly voted for it. Unfortunately, the two former members of the Labour Party, now independent, were persuaded to support the government and the bill was defeated.

Here Muldoon saw a chance to raise once again the bogey of the Soviet threat and rally the electorate to the cause of ANZUS. He announced that in future he would tolerate no dissent among his party's members on any measure that was a threat to the alliance. Among the National Party caucus (and a most unlikely member of it) was a young woman called Marilyn Waring. Bravely she defied Muldoon, an action requiring considerable courage in a caucus cowed by his overbearing personality. She said that she would not vote with her party on issues of defence or nuclear weapons. The Prime Minister saw opening before him an escape from the difficulties of government. Late on the night of 14 June, to the surprise and it seemed to me the dismay of his colleagues, Muldoon announced that there would be an immediate general election. The reason, he said, was that the government had nearly lost a vote that would put ANZUS in jeopardy. Because of the actions of Marilyn Waring, he could no longer be certain of his majority in Parliament.

I found it appalling that Muldoon chose to make Marilyn Waring a scapegoat. She stood true to her convictions and voted in accordance with them. Her actions had not led to the defeat of the government — the vote on the Nuclear Free New Zealand Bill showed plainly enough that there was no chance of the government's being defeated over ANZUS before the scheduled end of the parliamentary term. The timing of the general election was Muldoon's choosing. It was typical of him that he responded to a stand of principle by trying to make Marilyn Waring carry the burden of responsibility which belonged to him and him alone.

But justice was done. If there were any who actually believed that Muldoon's fears for ANZUS were the reason for calling the election, they kept their silence. I certainly didn't believe him, nor could I believe my luck. The 1984 general election was lost and won before the strutting started; Muldoon swayed into view and threw in the towel. Any Leader of the Opposition who could face the media that evening merely looking in command of himself would have to be the winner.

Having tried for the last year to present itself to the public as a kind of government in exile, the Labour Party wanted a campaign that would allow for a display of stability and authority. The less contentious the campaign, the better for the Labour Party. For that reason I wanted to focus on domestic issues and not on foreign policy. I knew that debates about what might happen in the event of the break-up of ANZUS could only create uncertainty in the mind of the electorate. The possibility that the alliance might be ended was, in 1984, beyond the comprehension of many people who had been brought up to believe that ANZUS was the guarantor of our sovereignty and that international economic intercourse was not feasible without the prophylactic of the treaty.

Whenever I had to address myself to our foreign policy I pushed what I knew was the popular side of the argument, the banning of nuclear weapons, and emphasised all the while the Labour Party's goodwill towards our friends and allies. I deplored only by implication the habit of some of them of arming themselves with nuclear weapons. The National Party, meanwhile, did not make as much of the nuclear issue as it could. Had the election been held when it was due to take place, in November 1984, I am sure that the National Party and the American embassy would have done more than they actually did to stir up uncertainty about the future of the ANZUS treaty. As it was, the embassy released in the middle of the campaign a speech by Assistant Secretary of State Paul Wolfowitz in which the usual warning was repeated. Making New Zealand nuclear-free by denying port access to American vessels would create operational difficulties in the ANZUS alliance.

The National Party was at its weakest when it tried to argue the merits and demerits of nuclear weapons. It drew a distinction between tactical nuclear weapons and strategic nuclear weapons, claiming that the American vessels which visited New Zealand were not carrying strategic weapons, and so were unlikely to attract the attention of those directing the Soviet arsenal. This argument was less than useful to them in face of the fear of nuclear energy that was driving much of the public opposition to the ship visits.

Muldoon's political instincts were sounder. The United States, he said in a television debate less than a week before election day, didn't bring nuclear weapons to New Zealand and hadn't done so for years. If this unabashed assertion, with its implicit repudiation of the doctrine of neither confirm nor deny, caused any disquiet at the American embassy, they weren't letting on.

In the end, the nuclear-free policy wasn't decisive. For reasons of domestic policy and good taste the National Party lost the election. Within an hour of the close of polling it was clear that Labour would win with a handsome majority. In the old hall at Mangere my celebrations were twice suspended. First came a call from Australian Prime Minister Bob Hawke. He put Secretary of State George Shultz on the line. Shultz was pleasant in his congratulations but reminded me that we had some things to talk about when he came to Wellington. Then came a call from the defeated Prime Minister. He conceded graciously, but added that he had an important matter he wanted to raise with me. 'You'll be hearing from me,' he said. Both these calls had an ominous edge to them. I went back to the party as if nothing had happened, but a *frisson* of apprehension prevented me from enjoying the rest of the evening.

The election took place on 14 July, a Saturday. Under New Zealand's constitution the new administration could not formally take office until nearly two weeks later. Convention had it that the outgoing administration would act, in the meantime, in accordance with the announced intentions and, if necessary, the direct instructions of the incoming government. But the conventions of the constitution did not allow for a wrecker like Muldoon. In his book, convention was there to be flouted. On the Sunday, he threatened economic crisis. On the Monday, he refused my instruction to devalue the currency. With the foreign exchange market suspended, the crisis was real enough. New Zealand was on the brink of defaulting on its international debt.

In the middle of this economic mayhem the last thing I

wanted or needed was a meeting with George Shultz, but the Secretary of State was on his way to New Zealand. Having pledged so much goodwill towards our American ally, I agreed to his request for a meeting. To stress the new government's warm feelings, I decided to be on hand to welcome him.

Shultz was coming to New Zealand for a totally spurious purpose. He was the American representative at the meeting of the ANZUS council. The extraordinary decision had been taken to go ahead with the meeting set down for Wellington on 16–17 July, two days after the general election. Almost from the start of the campaign the opinion polling showed that the National Party was likely to lose. The timing of the council meeting meant that New Zealand would be represented, not by the new Labour administration, but by members of the defeated government.

The ANZUS council was made up of the foreign ministers of the three member countries of the ANZUS alliance. It was set up by Article VII of the treaty 'to consider matters concerning the implementation of this Treaty'. Its meeting in 1984 seemed to me to be a calculated attempt to embarrass the new Labour Government by ramming it down the throat of the public that New Zealand could not remain in ANZUS unless we gave unimpeded access to visiting American naval vessels. In Australia on his way to New Zealand, Shultz seemed optimistic. He expressed the hope that Labour's victory in New Zealand would pose no greater threat to the ANZUS alliance than had Labour's win in Australia the year before. He was, after all, in the right place to express such a view. The Australian Labour Government's examination of its alliance obligations had concluded much to the satisfaction of the United States. Shultz was hardly being far-fetched in speculating that New Zealand's intention to renegotiate ANZUS might have the same result.

The day after the election I met Shultz at the military airbase of Ohakea, a few minutes' flying time north of Wellington. Here a Boeing 707 carrying any United States emissary had to land; the runway at Wellington airport wasn't long enough. I arrived from Auckland in an RNZAF aircraft that would carry the Secretary and me on the hop to Wellington.

I greeted Shultz and we took off through the blustery night

to Wellington. A southerly tore the air and we bounced about the sky. Through the turbulence we descended, saying hardly a word. I wished the flight ended. What Shultz thought he gave no sign of.

The next morning a strange justification for the holding of the council meeting made its appearance. What were described as 'senior sources' in the capital were reported as seeing some advantage in the meeting of the ANZUS council in the absence of any representatives of the incoming government. In effect, those anonymous sources claimed, the new government had won a year's grace to attune its policies to those of its allies. I didn't know whether those sources were New Zealand or American, but their condescension made me burn with resentment. They seemed to confirm what I already suspected. Shultz couldn't be in Wellington with the expectation of any serious negotiation with the new Labour administration. By all appearances, he had come to the council meeting not to talk to me or to the representatives of the defeated government but to talk to the New Zealand public.

At his press conference after the release of the ANZUS council communiqué, Shultz was scornful of the Labour Party's announced aim of broadening the ANZUS alliance into wider forms of economic and cultural co-operation. ANZUS, he declared, was a security agreement. That was the extent of it and that was the sum and substance of it. There was nothing to negotiate in the ANZUS treaty. He told his audience that he'd be happy to discuss the alliance with the new government, but he was certain that a thorough examination of it would lead the New Zealand Government to the same conclusion that the Australian Government reached the year before. He wondered out loud what kind of alliance it was if the military forces of the countries involved were not able to be in contact with each other. With these words he reminded his listeners of the likely price of the nuclear-free policy. But it seemed that the Secretary of State had not come to exact the price that very day; it was his understanding, he said, that no ship visits were planned for six months.

His manner was so pleasant, his temperament so equable,

that his words bore no taint of the overbearing. But the ANZUS communiqué left a bad taste in my mouth. 'Access by allied ships and aircraft,' it intoned, 'to the airfields and ports of the ANZUS members is reaffirmed as essential to the continuing effectiveness of the alliance.' Shultz's manner seemed at odds with his compliance in the intellectual dishonesty which allowed the representatives of a defeated government to put their country's name to a document that all who signed it knew did not represent the views of the country's future government.

It was arranged that Shultz would call on me when the council meeting was finished. Into my office on the ramshackle top floor of the old Parliament Buildings, where the Labour Party was quartered, came a volley of security officers. This was the advance guard. With them came the bomb dog. In a country where prime ministers can and do walk unaccompanied down the street unmolested by anything more threatening than offensive gestures, this was a revelation. A thorough search ensued. My office being declared safe, I went to meet my visitor at the early-model lift, which groaned upwards with its cargo of Shultz, officials and bodyguards.

We said to each other in private much less than we'd been saying to all the world in public. Shultz told me about the ANZUS council meeting; it was a success, he said. He talked about the communiqué and its reference to nuclear vessels being an important component of New Zealand's defence.

'That's not consistent with our policy,' I said.

'That's something we'll have to work through,' Shultz replied.

I stressed to him that this was really not the time to talk rationally about the issue, and he freely acknowledged this. It would have been hard for him to disagree. A battalion of the news media laid siege to my office. Some no doubt were attracted by Shultz's presence. Most were in pursuit of the politicians and bureaucrats who were trying to get the country out of the financial hole Muldoon had dug for us. Shultz was perfectly sensible about it. I told him that I hoped to be at the meeting of the United Nations General Assembly in September and that we would have the chance to meet again. We didn't get into any argument about our respective policies. I asked him about the extent of American concern about the nuclear-free policy, and the likely American response. He assured me that the United States would not put economic pressure on New Zealand to change its foreign policy. The impression he gave me was that he saw in American policy sufficient integrity and authority to make it capable of standing on its merits unfortified by threats of sanctions, or restrictions on New Zealand imports.

If Shultz had it in mind at that meeting that New Zealand would eventually change its nuclear-free policy, the thought hung in the air unspoken. He struck me as careful and cautious. I assumed that he was confident that if he left the New Zealand Government to a quiet period of reflection, it would be plain to us that we should get back to business as usual in the ANZUS alliance. But he did not say so, and I had no reason to take exception to anything he did say. Out of my office pressed the phalanx of security officers, through the scrum of reporters. Out of the mouths of Shultz and myself emerged platitudes. We will work together co-operatively across the board, he said. There will be continuing association and dialogue, I replied. These words, trite as they were, perfectly reflected my hopes. If we were going to make New Zealand nuclear-free without provoking a crisis in the ANZUS alliance, with all the electoral uncertainties that would pose, then the governments of New Zealand and the United States had to talk. The meeting with Shultz had left me exactly where I was before it started. I was at odds with the Americans but still hopeful that, in spite of the public posturing of the ANZUS council meeting, some honourable settlement would be found.

Shultz, of course, was not the only foreign minister in town for the council meeting. I met Bill Hayden, his Australian equivalent, at the Australian High Commissioner's residence on the fashionable heights above Wellington Harbour.

Hayden was once leader of the Australian Labour Party when it was in Opposition, and in that capacity supported the Labour Premier of Victoria in an attempt to ban nuclear-armed and -powered vessels from the ports of that state. He was then strong-armed, both by the conservative faction of the Australian Labour Party and by officials of the American administration, into accepting that ship visits should continue on the usual basis. In other words, the Americans would neither confirm nor deny the presence of nuclear weapons on vessels visiting Australia, and the Australian public would, as usual, be left to wonder if any particular vessel was carrying nuclear weapons. Hayden's embarrassing climb-down helped undermine his position as party leader and prepared the way for the leadership of Bob Hawke.

But he didn't take the chance to remind me of his experiences. Instead we passed a pleasant evening, warm in our feelings of recent electoral achievement. He told me that the Australians would be playing no part in the discussions between New Zealand and the United States about ANZUS. Australia would not be heavying New Zealand, he suggested. New Zealand and the United States had to work their relationship through by themselves. He told me that he hoped that our differences could be resolved, but that Australia wasn't going to be carrying any messages. Bill Hayden always meant what he said at the time he said it.

On 26 July the new Labour Government was sworn into office. The nuclear-free policy was easily put into force. By international law, no foreign military vessel or aircraft could enter New Zealand without the permission of the government. Once Labour became the government, permission to enter New Zealand would not be granted to craft that were nuclear-powered or were carrying nuclear weapons. In the simplicity of its implementation, I don't suppose the Labour Government ever made an easier decision. If its implications were clouded by uncertainty, I wasn't unsettled by the prospect. Nor were my colleagues in the new government. In the mild southern winter of July 1984 we were a happy bunch. Everything seemed possible to us.

Labour Prime Ministers do not choose their Cabinet. Ministers are elected from among their number by members of the party's parliamentary caucus. Several of the new Cabinet were strong in their support of the nuclear-free movement. Others were not so closely identified with it, but there was no obvious constituency for a reversal of the policy. In the caucus as a whole there was a large majority in favour of the policy. Outside Parliament, the Labour Party's rank-and-file membership was disappointed only in what it felt to be the unnecessary moderation and restraint of the government. At the party's annual conference in September, the usual resolution was passed calling for New Zealand's withdrawal from all military alliances. For good measure, the delegates insisted that the Americans be required to close down their Antarctic support base at Harewood in Christchurch. Be that as it may, no observer, in the American embassy or the State Department, could possibly have mistaken the strength of feeling in the Labour Party in support of the nuclear-free policy. It was in the fabric of the party now.

I was sure in my own mind that, as had happened before, the Labour Party's long-winded constitutional processes would filter out any reference to withdrawal from alliances long before we wrote the party's platform for the 1987 general election. I was equally sure that a continuing commitment to ANZUS was an electoral necessity. Quite apart from what I knew about popular support for the American connection, I had campaigned in 1984 on the insistence that we could and would stay in the alliance. I couldn't abandon it without a struggle. In spite of Shultz's assurances, I could not be certain about the extent of the American reaction to New Zealand's insistence on maintaining its nuclear-free policy. If there was going to be unpleasantness, it was essential that the United States was seen to be starting it.

I thought that if I kept stressing the government's intention to take an active part in ANZUS at the level of conventional armaments, then I had a fair chance of winning the battle for public opinion inside New Zealand. If the Americans, satisfied that New Zealand could not be swayed into changing its policy, would agree to some form of military co-operation inside the ANZUS framework, our policy would be vindicated. That was the best possible outcome for us. If the worst happened and the Americans simply refused to accommodate us, then at least, in seeking some agreement, I had made an effort.

Nothing now suggested the Americans were ready to come to terms with nuclear-free New Zealand. I proposed, for instance, to appoint Bill Rowling our ambassador to Washington, confident that he would faithfully represent the nuclear-free policy. I soon learned from the Ministry of Foreign Affairs that the Americans took exception to the appointment, precisely because Rowling was so closely identified with the policy. It grated on me to be told that the Americans did not like my choice. Bill Rowling was a former Prime Minister. His reputation had risen above politics. The American attitude struck me as petulant and small-minded. It was not a promising overture to my meeting with George Shultz. Late in September I left for New York, our little force headed squarely in the direction of the immovable object.

The immovable object was, as before, amiable. The meeting assumed, from my perspective, the atmosphere of an academic contest. I put New Zealand's argument: we did not wish to be under the American nuclear umbrella. We did not wish to be defended by nuclear weapons. To make this plain to all the world, we would not have nuclear weapons in our ports. This was a proper form of arms limitation, but it was not an abandonment of our responsibilities as an ally. It was still perfectly possible for New Zealand and the United States to co-operate militarily at a lower level. I revisited the arguments for co-operation, which, seen from the bottom of the South Pacific, were so convincing.

The response was uncompromising. Shultz, his voice even, was forthright in his exposition of the American position. The projection of American power in the Pacific was not to be divided. Conventional forces could not be separated out from the nuclear deterrent. The United States would not abandon its policy of neither confirming nor denying the presence of nuclear weapons on its vessels. It would make no exception for New Zealand. It was incumbent on an ally to accept the visits of American vessels. This was, according to Shultz, a small price to pay for the protection provided by the United States. He did not say that unless New Zealand accepted American ship visits there would be an end to ANZUS. Instead, he started to talk about a future beyond ANZUS; he hoped New Zealand and the United States would maintain their friendly relations whatever happened to their military alliance.

We had talked about plans for visits by United States Navy vessels in 1985, hedging around the unspoken American hope that some further delay in the resumption of the visits would give the New Zealand Government the chance to change its policy. Now I told Shultz that it would be wrong to let things drift. I said that the issue should be put to the test and that there should not be a continued period of stand-off. He agreed. To settle the matter, we decided that the Americans would do what they did in December every year, and submit to the New Zealand Government a request for clearance for next year's ship visits.

That was as close to detail as we got. Mostly we fenced. Shultz did not ask me to change the New Zealand Government's policy. I suppose those who make such changes must, if they are to keep their self-respect, be allowed to believe that they have reached the inevitable conclusion through their own processes of reasoning. Nor did I say to Shultz that I would rather abandon the ANZUS alliance than lose the nuclear-free policy; I pressed the case for both. I could not afford to have it recorded, and later, for certain, published, that I was the one who had summarily closed off the options.

Shultz raised the question of visits by nuclear-powered ships. He knew that I was on the record as willing to put the question of nuclear propulsion on a different footing from nuclear weapons. That was not the government's position, but it was still my own opinion. Now I was in a logical trap. I could hardly urge rationality on the United States while arguing for an approach I believed to be flawed. So I told Shultz that I would try once again to persuade the Labour Party to express its policy in terms that would not make the exclusion of nuclear-powered vessels a pillar of our foreign policy. I would argue that nuclear-powered vessels which were proven safe and were not carrying nuclear weapons should be allowed to visit.

Having given the undertaking, I wished to honour it. I hadn't forgotten the difficulties I got myself into in 1983. But I hoped that in government, having firmly excluded nuclear weapons, I might have more room to move. Not long after I returned to New Zealand I spoke in a radio interview about the possibility of reviewing the Labour Party's policy on nuclear propulsion. The reaction soon convinced me that I was heading for trouble. The news media saw only a backdown from the policy on which the government fought the election; the distinction between propulsion and weaponry, so compelling to me, was nothing to them when compared with the prospect of a reversal of policy. Enthusiasts inside the Labour Party were furious. It was obviously going to be an uphill struggle.

I didn't persist. The hard fact of it was that the enthusiasts were right and I was wrong. Most nuclear-powered vessels were nuclear-armed. Given the refusal of the United States to disclose the presence or absence of nuclear weapons, every nuclear-powered vessel might as well be treated as carrying nuclear weapons. In the end, I learnt that lesson.

5

Meeting the Threat

THE AMERICAN AMBASSADOR to New Zealand was H. Monroe Browne. He was a tall, rangy figure, who wore button-down collars and looked as if he dressed himself from a second-hand shop. Appearances were deceptive in his case, because the ambassador was known to be wealthy. He used to belong to the California 'kitchen Cabinet' of Governor Ronald Reagan. His old friend had sent him to New Zealand to spend the end of his working life in an embassy in a country where nothing much usually happened.

Monroe Browne was a man of profoundly simple philosophy. In his view, the United States could do no wrong. Even more fervently did he believe that Ronald Reagan could do no wrong. It was not, for the ambassador, a triumph of intellectual gymnastics to arrive at this position; it was the creed by which he lived. There was no room in his mind for any notion that others might disagree with his government, and he had no sympathy for those who did. I don't think he ever really understood New Zealand. He certainly couldn't get to grips with the concept of a Labour Government.

In person, Monroe Brown was affable. He was so genuine I couldn't help liking him even when totally at odds with his view. He certainly was industrious, but he was hampered by his inability to hear a great deal of what was said to him. In conversation, he would often lean forward, hand cupping his right ear, and say, 'Pardon? What was that you said, Mister Prime Minister, David, sir?' He always came to our meetings in the company of a foreign service officer from the embassy, who held up the American side of the discussion if the ambassador seemed to be faltering.

The ambassador and I met only for set-piece talks. Most of our negotiations with the United States were not carried on by elected politicians, or by the appointees of politicians, but by senior officials who remained at their posts whatever the colour of the government. There used to be a myth that this permanent corps of officials was politically neutral, and perhaps they were in the limited sense that they did not openly favour one government over another. But the myth didn't take account of the fact that every body of people develops its own corporate culture, with its own goals and values. These cannot in the nature of things be neutral, and they may be at odds with the values and goals of the elected government. So it proved with the nuclear-free policy. For the most part, the government had to pursue the policy against the advice of its permanent officials.

Chief among these was Mervyn Norrish, the Secretary of Foreign Affairs. He was head of the diplomatic service and the government's senior adviser on foreign policy. Tall, somewhat stooped, gaunt, he was an impressive figure in his tailored suits and his shirts with monogram on the pocket. Like H. Monroe Browne, he was hampered by deafness. Sometimes, when talking to both of them, I would hear 'Pardon?' in unison. Norrish was a model of rectitude. He never let personal feeling intrude on the official advice he was giving. I certainly trusted him to carry out my instructions faithfully, even when those instructions were at odds with the advice he had given me. It must have been very trying for him to be cast into the diplomatic maelstrom at the climax of his career, but he sacrificed neither dignity nor honesty in the discharge of his reponsibilities.

One or two senior officials of the Ministry of Foreign Affairs I counted on to represent the nuclear-free policy with some enthusiasm, but at other levels there was active resistance. A junior official in our embassy in Washington once told New Zealand journalists that I had asked to see some American officials and had been refused. This was a snub and worth reporting if true, but it wasn't. Such guerrilla interference was a continuing annoyance.

The sum of the advice I received from the ministry (and this is putting very crudely what was rehearsed endlessly) was that pursuing our nuclear-free policy would lead to a breakdown in New Zealand's relationship with the United States. It would have an unfortunate effect on our security arrangements. It could have severe economic consequences if the United States took action against New Zealand's trade. I don't remember ever receiving advice that I should abandon the policy. But I was forever the recipient of proposals for its modification, even to the point of its extinction.

Nor do I remember receiving a word of advice that could be taken as support for the nuclear-free policy. Instead, I was constantly told that good things flowed from our dependence on the United States, and that any attempt by New Zealand to alter the nature of the relationship was fraught with risk. New Zealand's acceptance of the nuclear deterrent was the price we had to pay for the assurance of a warm relationship.

My officials missed no chance to remind me of the harm they believed the nuclear-free policy was already doing. The government had resurrected the proposal to create a nuclear-free zone in the South Pacific. Now officials warned us that we 'should also be aware that the maintenance by New Zealand of a total and unqualified ban on port visits by vessels that are either nuclear-powered or capable of carrying nuclear weapons may constitute a significant complication in the realisation of an effective regional zone. Whatever we might say in explanation of our policy, the United States and Australia . . . are likely to suspect that New Zealand's long-term goal is to have that policy adopted region-wide and incorporated in a regional zone.' In other words, we shouldn't run the risk of hinting that the nuclear-free zone should actually be nuclear-free.

The officials who advised against the nuclear-free policy were not unpatriotic. They served their country's best interests as they saw them. They had examined the options open to us, and concluded that our greatest advantage lay in continued deference to the wishes of the United States. Some part of their feeling must have come from the lifestyle of the foreign service officer. The ease of our relationship with the United States had a certain enchantment for them. Our officials used to place a lot of value on what they called the 'influence' on the United States which the ANZUS treaty gave New Zealand, like the mouse whispering in the ear of the elephant. I didn't think they could possibly be serious; what is now on public record about the ANZUS meetings predictably suggests that the Americans did the talking and our side did the listening. Nonetheless, the thought of a disruption of this cosy arrangement set them aflutter. The phenomenon wasn't limited to our relationship with the United States. I read many reports from posts abroad in which New Zealand's interests were subordinated to the easier course, which would allow our diplomats to enjoy a continuing good relationship with the host government. This probably happens in any foreign service, but it made for trouble when the nuclear-free policy put ANZUS in jeopardy.

The Ministry of Foreign Affairs offered a restrained commentary on the government's policy compared with the clamour that arose from our armed forces. New Zealand had a tiny navy made up of four second-hand frigates and some support ships. The air force had some fighters, not state-of-the-art, and some transport aircraft. The army was mostly infantry, its fighting force spearheaded by a battalion stationed at Singapore to repel the Soviet threat before it got too close for comfort. Our armed forces weren't in the least capable of defending the country, but then nobody really expected them to. The forces, essentially, were a fragment of somebody else's. They were made to fight alongside other, larger forces, and didn't make sense in any other context. ANZUS was all in all to them. Outside the alliance, there wasn't much reason for their existence.

This was reflected in military attitudes to the nuclear-free policy. The Chief of Defence Staff was Ewan Jamieson. Unlike H. Monroe Browne, he had formed his view of the value of nuclear deterrence after examining the available range of intellectual options. In the early days of the new government, and perhaps in anticipation of a change of heart, he was supportive. He would crisply tell me where the government was going wrong, but just as firmly tell me how he proposed to put our wishes into practice. Later on, a little edginess crept in. When he retired he took to the lecture circuit, warning all who would listen of the irresponsibility of the nuclear-free policy and the dangers it posed to the defence of the West.

The government's chief civilian adviser on military matters

was Denis McLean, the Secretary of Defence. He was in an awkward position when the Labour Government took office, being on the public record as a supporter of visits by nuclear ships to New Zealand ports. He got around this awkwardness by ignoring it, choosing instead the path of resistance. In person he was a tall, lugubrious, large-footed individual, patrician in appearance, stamped with all the niceties of scholarship and manners. His distaste for the nuclear-free policy he scarcely took the trouble to conceal. I did everything I could think of to get McLean out of his job (not a straightforward proposition, since politicians were not supposed to meddle in bureaucratic appointments). It was within my power to make him an ambassador, but he refused to go anywhere but London, a plum I was not disposed to give him. I tried to get him appointed to the Department of Conservation, where he could indulge his harmless interest in flora and fauna, but he wouldn't budge.

The problem wasn't his opinions; I could accept his advice or leave it alone as suited me. The real difficulty was the comfort his presence gave to the disaffected in the Ministry of Defence and the armed forces. Of these, there were many. One way they had of showing their annoyance with the government was to leak stories to the news media complaining about reductions in military efficiency or falling morale among the armed forces. Reports filtered back to me of defence attachés overseas who spoke disparagingly of the government and its policy. Senior officers in unguarded moments on the cocktail circuit uttered words of bitter condemnation of the government's conduct.

For the most part these martial mutterings were no more than a nuisance. The defence establishment was neither large nor powerful. There were no large defence-based industries, and few lucrative contracts to be let. Professional soldiering wasn't an occupation with a lot of status. Although the armed forces had polished bureaucratic skills when it came to defending their perks and privileges, as a political lobby group they didn't have a lot of clout.

I never really understood the military. My father used to tell me about his time as a medical officer in the army. He was stationed far in the north of New Zealand, preparing with his fellow soldiers for the ever-receding possibility of Japanese invasion. He wore at all times the tin helmet with which he had been issued, reasoning that its wear must be essential or the army would not have been so ridiculous as to supply him with it. When saluted by the soldiery, he raised it politely.

The advice I got from defence headquarters had a lot in common with the messages that came to me, directly and indirectly, from the American embassy. The Americans, of course, put it a lot more bluntly. Nuclear-free New Zealand, they told me, was a breach in the Western solidarity on which the fate of the world depended. The two great superpowers were staring each other out and neither dared blink. New Zealand was an itch under the American eyelid. We were a signal to the Russians that the West's unity of purpose was splintering, its determination weakening, its commitment to nuclear deterrence faltering. If the Russians detected any sign that the West was backing away from its willingness to threaten the use of nuclear weapons, they would never be forced into arms reductions.

What they were really worried about was the example we could offer to others. They were straightforward in their concern about Japan. Under its constitution, Japan was formally nuclear-free; the Japanese were denied the manufacture or possession of nuclear weapons, nor were nuclear weapons to be brought there. From time to time, foreign warships visited Japanese ports on the understanding that their owners respected Japan's wish to be free of nuclear weapons. The official Japanese view would have it that the United States had not asked for permission to bring in nuclear weapons, and so it must follow that nuclear weapons were not being brought in. Such reasoning fitted perfectly with the American refusal to confirm or deny the presence of nuclear weapons on their vessels, but required, of course, a certain amount of hypocrisy on the part of the Japanese authorities. I didn't meet anyone with any knowledge of American military strategy, and who didn't have a vested interest in preserving American access, who contended that United States vessels invariably unloaded their nuclear weapons before heading for Japanese waters.

The Americans can't have been worried that New Zealand's

nuclear-free policy might have influenced the Japanese Government. Japanese administrators were of the same mind as they were. In January 1985, Prime Minister Nakasone paid a brief visit to New Zealand. I thought he might mention the nuclear-free policy, but he carefully avoided it. He invited me to visit Japan and I accepted, but somehow or other a time for the visit could never be made. The word from our embassy in Tokyo was that the Japanese Government was concerned that I might be an evangelist for a nuclear-free policy that was genuinely nuclear-free. If official Japan was willing to turn a blind eye, the Japanese public at large might not be.

The Americans were also anxious about another possible threat to Japan's official complacency. After my meeting with Shultz, the United States and New Zealand started to talk about a visit here by an American naval vessel. New Zealand, of course, had asked that the ship be nuclear-free. The Americans, equally, would neither confirm nor deny that any named vessel was nuclear-free, but as a token of good faith they offered to bring a vessel to New Zealand directly from Japanese waters. If New Zealand rejected the vessel, the spell that bound Japan might be broken. Ersatz nuclear-free policies in the end satisfied only if all concerned conspired to make believe that the limitation was real.

The Americans were also perturbed by the resistance to nuclear entanglements which showed itself every now and again in countries like Spain, Greece and Denmark. At the same time, they were mounting a push to intensify their nuclear deployment in Europe. They were troubled by public opposition to Cruise missile installations. Against this background, their message to me was pointed. New Zealand couldn't be allowed to get away with it. If we did, other countries might start to think that they could get away with it too.

When he was in a mood to be positive, H. Monroe Browne never tired of quoting the Australian example. Here was a government that had lived up to its responsibilities. Governments, the ambassador told me, did not exist to surrender to public opinion; they were put there to stand up to it. Browne's offsider from the State Department always made a point of ramming home this message. It wasn't just my task to call my country to order: it was my duty to do it. When I said that the nuclear-free policy had been arrived at democratically, not by me alone but by the party I represented, the ambassador or his offsider would look aghast at what they plainly thought was submission to the mob. If I pointed out that the nuclear-free policy was already widely popular, accusations would fly that the government was misrepresenting public opinion in an attempt to excuse its behaviour. Monroe Browne and his colleagues would not let pass unmolested any argument that suggested that the fate of nuclear-free New Zealand was other than mine, mine alone, to determine.

I think, looking back now on what happened, that their arguments had a double edge to them. There was, I suppose, an appeal to the vanity no politician lacks. The burden of what they were saying, after all, was that the great United States was relying on me, me alone, to turn the trick for them. If I succeeded, I could certainly count on brass bands on the White House lawn and photo opportunities with the President. The other side of the argument was that if I didn't turn the country around, then I was the person who had torn the fabric of Western security. I was the comforter of the Soviet Union. I was singlehandedly undermining Western resistance to the evil empire. I actually used to brood on that a lot. I felt instinctively that it was nonsense, but you can sometimes stop believing in yourself when the people you thought were on your side are saying you're letting them down.

The Americans were carefully calculating in the manner of their presentation. I was told that the United States was going to take me into its confidence. Secrets would be whispered which I alone would hear. Grave facts about the Soviet threat to Western security would be disclosed to me; I alone in the government was to be trusted with them. What was told to me was not to be discussed with my staff or my colleagues. The information was so sensitive I was not to put security at risk by talking about it.

I agreed to hear what they had to say, on their terms. I could hardly be a credible opponent of nuclear deterrence if I wasn't willing to listen to the arguments in favour of it and reply to them in a thoughtful way. I think now that the insistence on secrecy was part of a conscious attempt to isolate. When evidence of the Soviet Union's offensive capabilities and aggressive intentions was put in front of me, I would be barred by my own undertakings from discussing what I'd seen with others. Left to myself to imagine the extent of the risks I was running, I might be more open to suggestion that I could discharge my responsibilities to the Western alliance by ditching the nuclear-free policy.

The intermediary between me and the American intelligence community was Gerald Hensley, who in 1984 was head of the Prime Minister's Department and our External Intelligence Bureau. The messages he brought me were sometimes disconcerting. One day when New Zealand and the United States were still on speaking terms he padded into my office to bring me sombre news. He had just been informed, he said, that the American early-warning system had picked up Soviet missiles heading for the continental United States. It looked, he continued, as if nuclear war had finally happened. In twenty minutes we'd know for certain. There wasn't anything I could do about it, and, not wishing to start a panic among the staff, I sat signing letters while the minutes dragged past. Hensley came back. False alarm. It must have been a fault in the electronics or a close formation of Canadian geese. Nothing the Americans did intentionally ever frightened me as much as the fright they gave me inadvertently.

My office was close to the top floor of a circular building known to all as the Beehive. (An eminent British architect drew it one day on a table napkin.) One day I returned to my office to find it darkened. Shutters of hardboard had been put up against the windows. I learned afterwards that those unsightly novelties had been put in place to reduce the possibility that a microphone aimed at the windows would be able to pick up conversations inside. The only foreign embassy I could see from my office belonged to the United States, but the two visitors Hensley now introduced to me were cautious. These two arrivals from Washington were military in bearing and well groomed in their business suits. A screen had been put up, a projector was waiting. My guests were going to show me some pictures.

Across the screen, satellite imagery moved. Here were naval bases. Here were massive build-ups of armaments on the shores of the Soviet Union. Here were forests dotted with missile silos. I watched as the meaning of each scene, and the threat each contained, was explained to me. It was endless. It was worse than Mick Connelly talking about the police. I sat back in my seat. The room was dark. The voices droned. The pictures kept coming. I fell asleep. If the Americans were annoyed by this unintentional impoliteness, they concealed it. It should be noted here that intelligence officers, or at least the kind who carry out their work behind a desk and not in the field, always travel round in pairs. They are easily annoyed if you ask them, 'Which one of you two isn't to be trusted?'

The pictures didn't change my view of nuclear deterrence. Whatever the Soviets were up to, there wasn't anything New Zealand could do about it. It was idle to imagine that occasional visits by American ships to our ports would have any impact on thinking at the Soviet high command. Our ban on nuclear weapons, after all, would not reduce by one the number of nuclear weapons in the world. Nor was I moved by the thought that the Americans might use the threat of nuclear weapons to deter a conventional attack on New Zealand, if any such attack should ever be in the offing. A nuclear exchange could not be contained. Every scenario suggested that force would be met with equal force until the entire arsenal was unleashed and the planet destroyed. It was madness for New Zealand to be defended by weapons whose use would lead to our extinction. Far better not to invite nuclear retaliation. I couldn't believe in any case that the United States would destroy itself to defend its ally — they'd have to be mad.

The Americans got New Zealand wrong when they put so much emphasis on the doctrines of deterrence. People don't think that nuclear weapons are the answer to anything unless they're very frightened. The only group in New Zealand who swallowed deterrence whole were the military, and they were frightened about losing their cosy relationship with the United States. In the absence of an imminent threat, the Americans were talking about deterrence in a vacuum. Apologists for the

United States found it heavy going when they tried to drum up public support for the nuclear deterrent. One who made the attempt was Sir Robert Muldoon's successor as leader of the National Party. The most effective phrase I ever coined in politics I used against him in a parliamentary debate: he wanted to 'snuggle up to the bomb'. In New Zealand, the image worked. A nuclear defence was far more frightening than any threat.

The fact of it was that American insistence that New Zealand let itself be defended by nuclear weapons didn't have anything to do with military strategy; it was political in origin. If New Zealand didn't carry its share of the burden, countries where the deployment of nuclear weapons really mattered might think that they too could follow New Zealand's example. It was like the American policy of neither confirming nor denying the presence of nuclear weapons on their ships and aircraft: it had little to do with military secrecy and much to do with pacifying public opposition to nuclear deterrence. In New Zealand, no amount of computer-enhanced satellite imagery could make you think otherwise.

I can't remember the Cabinet ever sitting down to discuss the pros and cons of nuclear deterrence, but then I didn't ask them to. Nobody wanted to debate it. We talked a lot about the nuclear-free policy, but nobody questioned its value as a means of arms limitation or wondered about the effect it was having on the Russians. We discussed the policy in terms of its likely political and economic costs. We knew, without saying, what the political cost would be to us of abandoning it. What was worrying us at the end of 1984 was the cost the Americans might exact from us for keeping it.

The price of our policy we heard directly from the American ambassador, and indirectly from our own officials in the Ministry of Foreign Affairs and Ministry of Defence. In the first place, all military co-operation between the United States and New Zealand under the ANZUS alliance would be suspended. The Americans did not bother to spell out what this meant. They happily left it to defence headquarters to tell the government how this action would curtail training, lower morale and stop recruiting. Papers piled up on my desk, each protesting the inability of our armed forces to function effectively unless under the wing of the United States. The new government was hardly sworn into office when a report arrived from the Ministry of Defence which said, 'We lack air defence of airfield and ground forces, heavy artillery, satellite communications systems and over-the-horizon radar. There is no effective interceptor-fighter capacity in the air and no surface-to-surface nor air-to-surface guided missiles for any of the three services.' The list went on for some while longer before the ministry concluded that many of its wants were beyond our resources to remedy and that we had no option but to look to our allies. The United States was our major supplier of military equipment, which it sold to us at cost. And, the ministry reported, we relied on the intelligence information provided to us by our allies. If I ever needed persuading that our armed forces were hopelessly locked into the American perspective, those papers would have done it.

Here again, I think the Americans miscalculated in relying so much on the diplomatic and military establishment to put pressure on the government. At the height of the Cold War it might have worked. In 1984, most of Labour's Cabinet and caucus were of a temperament which found military posturing ridiculous rather than intimidating. Our pride in our country was not bound up with its military prowess. In New Zealand the armed forces did not have the standing which allowed them to dictate to governments. More than a generation gap was at work. Two older members of the Cabinet, both of whom served in the Second World War, were fierce supporters of the anti-nuclear policy, sometimes disconcertingly so. One of them made memorable the lunch the Cabinet gave when H. Monroe Browne retired. I was determined to be pleasant to the departing ambassador, notwithstanding another recent, and public, assault on the policy of his hosts, but Bob Tizard thought differently. He began with an old chestnut, America's belated entry into the war in which he had served. 'We were in it two years before you,' he snarled. 'We don't want you here.' Monroe Browne looked startled. 'Every time you come here you just cock it up!' I'm not sure if the ambassador heard what he said, but then he didn't hear much of what was said to him in New Zealand.

I had no reason to think that the Americans ever went beyond simple argument in their campaign against the nuclear-free policy. Rumours abounded in Wellington that a CIA front was going to finance the National Party's next election campaign, but rumour was all that came of it. Gerald Hensley told me before I went on a trip to the United States and Europe that I shouldn't discuss confidential matters in hotel rooms, so I didn't. In a fit of enthusiasm, Gerald and I decided that hardboard shutters were too cumbersome and that my office should be fitted with curtains specially treated to defeat listening devices. Eighteen months after they were ordered they finally turned up. That was the extent of our security precautions. I didn't believe I needed any. The greatest security scare of 1984 was provided not by the Americans but by the Russians. To my official residence was delivered a samovar, the kind gift of the Soviet embassy. I eyed it suspiciously. Who knew what sensitive devices lurked within it? The Security Intelligence Service came and took it away. The samovar proved to be innocent of listening devices, but it was wrongly earthed and highly dangerous. It came back to me rendered safe for tea-making.

H. Monroe Browne never explicitly threatened economic sanctions against New Zealand; George Shultz had said that economic and security issues were separate and that trade sanctions were out of the question. The ambassador took a different tack. New Zealand sold products to the United States which were in direct competition with the products of the powerful American farm lobby. There was always pressure from American farmers to restrict agricultural imports, and sometimes the pressure led to protectionist legislation in Congress. The ambassador kept on pointing out that the administration didn't have a monopoly in protectionist initiatives. It was quite likely that legislators, for political or economic reasons, would take action against New Zealand. If that happened, and New Zealand insisted on pursuing its nuclear-free policy, the administration would not intervene on our behalf. If New Zealand was an ally, it would only be proper for the administration to tell the Congress that protectionist legislation would damage a small ally's interests. If we weren't an ally, too bad for us.

There was menace in this. If American legislators were incited to put greater restrictions on our farm exports, the economic damage would be severe. As a country we were used to finding our way through the intrigues of American farming politics. My difficulty now was that I had no way of telling how strong the signals might be from administration to Congress that it was open season on New Zealand. The ambassador of course, wasn't about to let me know. He smiled to himself. I smiled right back, but only on the surface. When it came to our export trade, I had a duty to be worried.

6

Confirm and Deny

WHEN I FINISHED my business in New York I gathered up the little cakes of soap and bottles of aftershave so lavishly provided by the hotel management and took them back to Wellington as souvenirs for my office staff. They were not heavy baggage. The real burden was my agreement with George Shultz that the United States would ask the New Zealand Government for access for American vessels to New Zealand ports.

No more than Shultz had asked me to water down our nuclear-free policy had I asked Shultz to make an exception to his government's persistent refusal to confirm or deny the presence of nuclear weapons on American vessels. The sphinx was not going to be winking at me. We would make our own assessment, from our own resources, of the armament of any visiting vessel. The Americans might say as little as they pleased. It was our judgment, not their silence, which would decide the issue.

No American official, at any level, ever told me that the ship the United States wanted to send to New Zealand was not armed with nuclear weapons. Nor did any American ever tell me that it was so armed. As far as I was concerned, American adherence to the policy of 'neither confirm nor deny' was absolute. H. Monroe Browne gave no assurances; I was entrusted with no confidences. Unlike the missile-studded steppes and forests of the Soviet Union, the secrets of the US Navy were not laid out in front of me.

For all their reticence, the Americans wanted the New Zealand Government, when it came to the crunch, to make a judgment in favour of a visit. I'm sure they weren't expecting a sudden about-face. It was never suggested, for instance, that the Americans would start off by asking to send a nuclear-powered ship. They must have calculated that such an obvious turnaround, made so suddenly, would have been politically unsustainable. As Assistant Secretary of State Paul Wolfowitz afterwards put it to our ambassador in Washington, there had to be sufficient political preparation of public and party opinion in New Zealand before an accommodation could be arrived at. In the short term the American aim was to shepherd the New Zealand Government into convincing itself that the ship on the horizon was compatible with the nuclear-free policy.

Left unspoken was the larger question of what kind of nuclear-free policy we were talking about. The Americans must have calculated that the New Zealand Government had a choice. We could be genuinely nuclear-free; or, like other governments, we could declare ourselves to be nuclear-free and invite our nuclear-armed allies to respect our policy. In other words, we would turn a blind eye to the occasional presence of nuclear weapons in our harbours.

The chance to resolve these questions was now on offer. Many months before, the three ANZUS partners had agreed to hold a naval exercise, set down for March 1985 and christened Sea Eagle. Several American vessels of assorted types and conditions would be taking part. One might easily visit New Zealand. We suggested to the Americans that they present us with a list of participating vessels and from among them we would select one to call on us. The United States refused. Allies were not allowed to pick and choose.

The Sea Eagle exercise remained the focus of the talks. It was made plain to me that the United States would not make a formal request for a visit unless the request was going to be accepted. I agreed to talk it through. Having made a public virtue out of my willingness to reach an understanding with the United States, it seemed to me that we had to keep going.

Most of the talking was done by the bureaucracy. The Chief of Defence Staff went to Hawaii to consult the American Pacific Command. The Secretary for Foreign Affairs went to Washington to sound out the State Department. Little serious negotiation took place between myself and Monroe Browne. We talked, but not productively. The ambassador's exasperation at my inability to accept the logic of deterrence grew more obvious. I in turn was aware of his bewilderment and certain of my inability to get through to him. If the government's official advisers couldn't reach an agreement with their American counterparts, the ambassador and I were never going to be able to do the job for them.

My advisers and I had one goal in common. We wanted to see an American naval vessel in a New Zealand port. Beyond that there was a critical difference. The diplomatic and defence establishments were willing to see the visit made on American terms. To secure American goodwill, they would accept a visit by a vessel that might or might not be nuclear-armed. I don't believe for a moment that the government's senior officials ever tried to lock us into an agreement by misleading me, or by misrepresenting me to the Americans. Where their inclinations showed was in the enthusiasm with which they set out to convince me that the Americans were going out of their way to send us a ship that conformed to our policy.

I learned that New Zealand's intelligence services had no means of telling which American warships were nuclear-armed, beyond the kind of information available to any reader of *Jane's Fighting Ships*. Some nuclear weapons systems could be scientifically detected, but we didn't have the technology. Any determined foreign power could locate the whereabouts of nuclear weapons through on-the-ground intelligence, but we didn't have spies in Pearl Harbour. Instead, my advisers drew their conclusions about naval armaments from the history of each vessel. Its deployment would tell the story. Some older vessels, although still capable of carrying nuclear weapons, were not routinely nuclear-armed. A modern nuclear-capable vessel returning from a long sea-patrol was almost certain to be nuclear-armed. A vessel sent from its home port to show the flag around the South Pacific was less likely to be nuclear-armed.

Not all of the scenarios put forward were entirely compelling. The mere fact that a ship came to New Zealand straight from Japanese waters was not in itself a guarantee that the vessel carried no nuclear armament. But my advisers told me that in some circumstances it was possible to be, if not entirely certain, almost entirely certain that a given vessel was not carrying nuclear weapons. More than that, they told me, the Americans were willing to send a vessel under conditions which would lead us naturally to that conclusion.

I saw no harm in the search for a satisfactory scenario. If circumstances led to the firm conclusion that a vessel was not nuclear-armed, it didn't worry me that the vessel was still capable at some time of carrying nuclear weapons. Some ships, for instance, were armed with the ASROC anti-submarine system, which could deliver a conventional weapon or a nuclear weapon. If the system was conventionally armed when the ship came to New Zealand, and no nuclear weapons were on board, it would be welcome. I didn't object to its latent nuclear capacity any more than I objected to New Zealand ships taking part in conventional naval exercises with American nuclear ships. In this I was at odds with many in the wider nuclear-free movement, who believed we should avoid the least taint of deterrence. My reasoning was that we couldn't pretend that the weapons didn't exist; what mattered was willingness to limit their deployment. I wanted the Americans to prove to us that they had the will to keep nuclear weapons out of our ports. I didn't care about the ship — if I knew it wasn't nuclear-armed when it arrived, I'd stand on the wharf with Monroe Browne to welcome it.

In one sense any ship that visited would have to speak for itself. The Americans would not allow me to say openly that a particular vessel passed the test of New Zealand's nuclear-free policy. This I learnt in October, when a military exercise called Triad was held in New Zealand. Fighter aircraft of the US Air Force took part, capable of delivering a nuclear warhead. It was, however, plain to observers that the aircraft that came to New Zealand for the exercise had left their nuclear weaponry at home. There was not the appearance of it, nor any of the precautions ordinarily associated with its presence. To me, the Triad exercise was a demonstration that the government's wish for low-level military co-operation within the ANZUS alliance was perfectly workable if the Americans allowed it. For their part, the Americans made it plain that they would take exception if I announced their aircraft to be here in compliance with New Zealand's policy. This they described as a 'characterisation' of their position. The practice of characterisation was forbidden.

The aircraft came here with all the ambiguity the United States could muster, but their presence proved of little concern to the public. Nobody seriously believed them to be armed with nuclear weapons. The nuclear-free policy was uncontested. Any ship that arrived here would have to meet the same standard of common sense.

The negotiations went on in deepest secrecy. I assumed that was what the Americans wanted. If we were unable to reach agreement on a vessel, they would hardly want it publicised. It also suited me. Among many in the Labour Party outside Parliament, and in the wider nuclear-free movement, there was certainty that I would do as the government in Australia had done and embrace deterrence. The smallest contact with the Americans was regarded with suspicion. I didn't want to leave myself open to American accusations of bad faith by trying to carry on the negotiations in the middle of a media circus.

The only two in the Cabinet in whom I confided were Deputy Prime Minister Geoffrey Palmer and the number three in the Cabinet, Mike Moore. Palmer was entitled, as deputy, to know what was happening. Moore, an advocate from way back of the nuclear-free policy, saw the impasse with the Americans as a challenge to his firmly held opinion that every problem was there to be solved. The others I did not consult; there wasn't any need. There was nothing to persuade them of. If the ship was plainly nuclear-free, Cabinet, caucus and I would rejoice that the Americans had agreed to co-operate in maintaining the alliance. If the ship was ambiguous, the government collectively would have to make the choice between the wrath of the United States and our nuclear-free policy.

In the middle of December 1984 a diplomatic note from the United States arrived in my office. It was notice of American intention to make an application for a ship visit in 1985. I sent back an acknowledgment. I was asked about the note at my next press conference and repeated what by now I could say in my sleep. Nuclear-armed and powered vessels would not be visiting New Zealand. I hoped that vessels of our allies might call at our ports after the Sea Eagle exercise. We would be making our own judgment of each vessel's armaments according to its type, its recent movements and the weapons system each carried. I exuded, I hoped, confidence. 'The American people and government, are, in my view, intelligent and aware, and they are not going to engage in some sort of needless, pointless, provocative incident,' I declared. 'We have had the utmost co-operation from the United States over this matter.' And so, my official advisers told me, we had.

There was nothing for me to do but wait until the American request was actually made. I received another message. This one came to me hand-delivered by the Australian High Commissioner in an envelope marked 'Top Secret' and sealed with wax on all four sides. Opened, it revealed a letter from Prime Minister Bob Hawke. If the Australian Government was once determined not to browbeat New Zealand, it had changed its mind. He began by telling me what I already knew. Australia was committed to ANZUS and believed that the treaty obliged it to accept visits from American naval vessels. He said that there would be strains in the relationship between the treaty partners if New Zealand insisted on special treatment. 'We cannot,' he wrote, 'accept as a permanent arrangement that the ANZUS alliance has a different meaning and entails different obligations for different members.' I read this as saying that New Zealand was not to disturb the delicate balance of Australian public opinion by failing to take up our share of the burden of deterrence.

Now it was January, when political life in New Zealand comes to a stop for the summer holidays. I was going to visit the Tokelau Islands. The Tokelaus are a New Zealand territory. There are three small islands which lie in the South Pacific about three hundred miles north of Western Samoa. Each island is home to roughly five hundred people. Most of the Tokelau people now live in New Zealand. There is no airstrip and no harbour. The only transport to and from the islands was by tramp steamer, which the New Zealand Government chartered to make the round trip from Apia, capital and principal port of Western Samoa.

Every year the leaders of the Tokelau people made the long journey to Wellington. In 1984 they came to see me to lodge their protest at what they felt to be undue pressure on the islanders by the decolonisation committee of the United Nations. They did not wish to have independence forced upon them and asked me to the Tokelaus to see for myself. No Prime Minister had ever been there and I was glad of the chance to go. As well as some staff and a few journalists of the hardier type, I was going to take my sons Roy, aged thirteen, and Byron, aged eleven. I planned to combine the business of the official visit with a family holiday.

It worried me not at all that a request for a ship visit might arrive in my absence. The American embassy would not be looking for publicity. If a proposal was received, it could wait till I got back.

Off we flew to Apia, where we boarded MV *Avondale*. This elderly craft, we were shortly to discover, had just donated its last two tonnes of fresh water to Samoa; gravity had proved stronger than the pressure of the Apia water supply. The *Avondale* should have been scrapped some years before. Whatever sea-keeping qualities it might once have had were now diminished and it made no pretence at comfort. For the next eight days it was our home. Four and a half knots was its greatest speed, but it got us to the Tokelaus.

At each island of the group the *Avondale* circled in the deep water off the barrier of the reef. Its passengers descended a ladder made of fraying rope with wooden rungs. We went ashore in little boats which shot through the water crashing over the coral reef.

Here was perfect tranquillity. We had no radio or telephone contact with the outside world. There were no motor vehicles, no radio, television or video. There was no cash economy. The people harvested the sea and the land and gave thanks to God for both. The maximum penalty for murder was a fine of two hundred dollars (this was the result of a drafting error in the penal code written for the islands in Wellington), but despite the licence, homicide was unknown.

I met the islanders and talked to their leaders. I played cricket for Argentina against England. Argentina triumphed. (There were sixty-four players on the Argentine side.) When it came time to leave we trekked before dawn across the last island we visited. Luggage on our heads, we waded the lagoon. By the light of hurricane lamps we stumbled through the palms fringing the shore, and in the shelter of the wreck of a Japanese trawler we waited for the sun. When it rose we saw the *Avondale*, waiting to take us to Pago Pago in American Samoa. There we would meet an RNZAF Boeing 727 that would fly me to Wellington in time for the next meeting of the Cabinet.

We wallowed on a heavy swell. *Avondale* seemed hardly to move. We were in mid-ocean when I was summoned to the radio room. The air howled with static and a message came through letter by letter, which I wrote down as each surged in and out of hearing. What it meant I found hard to work out. A request had been made for a visit; that I understood. More followed. My chief press secretary, Ross Vintiner, would meet me in Pago Pago to tell me about events at home. This was frustrating; Ross wouldn't go to Pago Pago for the sake of the journey. I was left to wonder what on earth had happened.

I willed the *Avondale* forward. The ancient craft resisted — it was now about as sick as most of its passengers. We were late at Pago Pago, where the Lieutenant Governor of American Samoa welcomed us with doughnuts. At last we were on the aircraft and I finally heard what had taken place in New Zealand in my absence. The American request for a ship visit had become, spectacularly, public knowledge. As far as I could tell, the folks at home knew everything but its name.

The ship the Americans wished to send us was USS *Buchanan*, an elderly destroyer of the Charles F Adams class. Apart from its two five-inch guns, it was armed with Harpoon anti-ship missiles and Standard anti-aircraft missiles. These were conventional weapons systems. It was also armed with the ASROC anti-submarine system, which could launch a missile carrying a nuclear depth-charge. But the ship was not invariably nuclear-armed. The *Buchanan* was the very model of ambiguity, but in the opinion of the government's official advisers, it was unlikely to be carrying nuclear weapons when it made its visit to New Zealand.

The news reports Ross Vintiner brought with him left no doubt about the dilemma now faced by the New Zealand Government. The story came from unnamed official sources in Washington by way of Australia. The *Sydney Morning Herald* had said that the United States had made a request for a ship visit. While the ship in question was not nuclear powered, a question mark hung over its armament. According to the *Herald*, the decision to go ahead with the request for a visit was designed to put New Zealand and its policy on the spot. It was a move intended, the newspaper reported, to put immediate pressure on the government.

A lot of politics is posturing. I don't know what made the State Department come over macho in the *Sydney Morning Herald*, but it did, and it sank the *Buchanan* with it. The assurances my officials had so carefully gathered as they worked through their scenarios with the Americans were now valueless, destroyed by the brutal assertion that the *Buchanan* visit would be a triumph of American nuclear policy.

It was commonplace in Australia for sensitive government documents to end up in the newspapers and there soon appeared in the press the letter I had received, all bound up with sealing wax, from Prime Minister Hawke. Who might have leaked it, nobody knew. I assumed the Australians wanted to jump on the bandwagon of the American triumph. It didn't help.

The news media hummed with speculation. The class of the vessel was correctly identified. Details of its weapons systems appeared in the newspapers. The tactical capabilities of the ASROC system were analysed. The ship might be nuclear-armed. It might not. Nobody could say for certain. The Chief of Defence Staff and the Secretary for Foreign Affairs understandably made no contribution to the public debate.

In my absence in the Tokelaus, Geoffrey Palmer was Acting Prime Minister. Assailed by the press gallery, Palmer turned to his guiding light. Procedure was his lodestar. Properly applied and duly followed it would lead inevitably to the correct result. He made it clear to journalists that the government was prepared to welcome a nuclear-capable ship, if we were satisfied on the evidence that at the time of its visit to New Zealand the ship was not carrying nuclear weapons. As he pointed out, we could not tell in broad terms what vessels were carrying nuclear weapons. But, he added, 'we have the capacity to decide in a specific instance whether, in the context of that visit by that vessel, it conforms with our policy or not'. The government would rely on the advice of its officials, who would rely in turn on their own technical expertise and what Palmer called 'New Zealand's analysis of the strategic situation in the South Pacific'.

In the overheated climate created by the Australian reporting, Palmer's observation that the government would be willing to welcome a nuclear-capable ship pushed alarm buttons in the nuclear-free movement. The anxiety of its members was, in the circumstances, understandable. They could not see how the New Zealand Government, from its own resources, could possibly be certain about the vessel's armament. And they were right. Palmer pored over the reports of the External Intelligence Bureau and the Ministry of Defence. They could not conclusively say if the *Buchanan* was nuclear-armed or not. The government's advisers acknowledged that some small chance remained that the destroyer did not comply with the government's policy. Palmer wrote out a recommendation to me that the visit not be accepted.

In the aircraft flying towards home I read Palmer's report. I also read the recommendation of the Ministry of Foreign Affairs that we allow the *Buchanan* to visit. The ministry's extensive consultations with American officials led it to conclude that the *Buchanan* was almost certainly not armed with nuclear weapons. It almost certainly complied with our policy and should in the ministry's view be admitted.

I scribbled an endorsement on Palmer's report. The *Buchanan* had lost its innocence. Near-certainty was not now enough for us. Whatever the truth about its armaments, its arrival in New Zealand would be seen as a surrender by the government. When it dropped anchor in our waters, it would hopelessly compromise us. Nuclear-free New Zealand was meaningless if the policy was tainted. Our stand against nuclear weapons would make no difference unless we were in a real sense distinguishable from governments that proclaimed themselves nuclear-free and then turned a blind eye to nuclear weapons. To be true to our policy and ourselves, the USS *Buchanan* had to be rejected.

The aircraft landed in Wellington and I drove straight to the Cabinet meeting. Our discussion of the *Buchanan*'s visit was short. Ministers took it for granted that the visit would not take place; we examined the written submission of the Ministry of Foreign Affairs and I put the recommendation that permission for the *Buchanan*'s visit not be given. There was general agreement. I added a rider. If the United States would neither confirm nor deny the presence of nuclear weapons on its vessels then New Zealand would only issue invitations to ships that were, indisputably, free of nuclear weapons. That ended it at Cabinet.

At the press conference I held after every Cabinet meeting, I stalled. I didn't want to tell the journalists that we'd rejected the American request since I hoped there was some slight possibility that the United States might withdraw its nomination of the *Buchanan* and substitute another, unambiguous ship. The matter was still under investigation, I said, the decision still to be made. Wishing to suggest to the journalists that the matter was not without the possibility of a satisfactory resolution, I observed that there were some vessels which were simply not capable of carrying nuclear weapons, and were known as such. Thus ended the long day that began on the rolling Pacific on the elderly *Avondale*.

The next day, H. Monroe Browne presented himself in my office. The mood was sombre. I told the ambassador that the visit of the *Buchanan* would not go ahead since the ship did not conform to our policy. The ambassador was not pleased; he told me that it was my responsibility to persuade the Cabinet of the virtue of the visit. This patronising assertion I found out of place. I pointed out that any possibility of a reasoned assessment of the *Buchanan*'s armaments had been utterly destroyed by publicity for which New Zealand was certainly not responsible.

The ambassador recited, as he had often done before, the likely consequences of our refusal. This I expected. I waited until the last of the lamenting for the demise of ANZUS had faded. Then I said that there might yet be a way out of our difficulty. The government would welcome the visit of an American vessel of the Oliver Hazard Perry class, a type universally understood to be armed only with conventional weaponry. Such a visit would demonstrate that military co-operation between New Zealand and the United States was, for all our differences over nuclear weapons, a practical possibility. The ambassador looked doubtful, his accompanying officials slightly less so. But he agreed to think this invitation over.

A few hours after the ambassador left my office, news reports announced that the government had asked the United States to send a ship of the Oliver Hazard Perry class. This was infuriating; if there had ever been any chance that the Americans would agree to substitute another ship for the *Buchanan*, there was an end to it. I had my office electronically swept for bugs, but there weren't any. I later heard it suggested that journalists worked out what was happening from the comments I made about demonstrably nuclear-free vessels at my press conference after Cabinet. I don't know the truth of that; in fact to this day I don't know what happened.

Monroe Browne was back in my office again the following day, his manner intense. He told me no substitute vessel would be put forward. I said that that was understandable in the light of what was in the newspapers. He said it was in the newspapers because I'd leaked it. No, I said, the leak hadn't come from us, I was sure of it. What was in the papers was there because his side had put it there.

'Well, anyway, it doesn't matter,' the ambassador spluttered. 'You're not going to get it, Mr Prime Minister, David, sir!'

The same day fifteen thousand people marched down the main street of Auckland to stiffen what they fancied was the government's failing resolve. 'If in doubt, keep it out!' they chanted.

On the next day the Labour Party's parliamentary caucus held its weekly meeting. I told the caucus in outline what had happened and said that the nuclear-free policy would be upheld. This met with applause. Hoping that the ambassador might not have the last word on the subject on the American side, and wanting to show willing, I wrote to Monroe Browne and formally extended an open invitation to the United States to send a vessel of a non-nuclear-capable class.

The State Department, meanwhile, was taking the offensive. A statement issued from Washington. 'The denial of port access,' it declared, 'would be a matter of grave concern which goes to the core of our mutual obligations as allies.' If New Zealand did not accept an American ship, military co-operation under ANZUS was in question. This theme was eagerly taken up by the government's parliamentary Opposition. America was not bluffing, the new leader of the National Party declared. The price of the nuclear-free policy was the end of ANZUS, he said. In this dismal refrain joined the country's editorial writers, by and large a conservative lot. One had us 'blundering out of ANZUS'. The whimsically named *New Zealand Herald* reminded its readers of the nobility of principle that underpinned the ANZUS treaty. Pointing the editorial finger squarely at the government, the *Herald* added that 'far too many people seem to be putting wishbone where backbone ought to be'. In New Zealand, fortunately, only their authors take editorials seriously.

On 3 February 1985 the American ambassador wrote to me with the stark advice that it was the *Buchanan* or nothing. This letter I took to Cabinet. The Cabinet had not changed its mind: the *Buchanan* was not coming. I wrote back to the ambassador telling him that New Zealand was unable from its own resources to determine if the *Buchanan* conformed with our policy, and for that reason we must decline its visit. I told him that our decision in no way diminished our commitment to the ANZUS alliance.

This was not how our allies saw it. Prime Minister Hawke announced that the Sea Eagle exercise was cancelled. A few hours later the United States announced that they were withdrawing from the already-cancelled exercise. Not for the first time I wondered how they'd get their act together if anything serious ever happened. Then came the irony. Bob Hawke had his own problems with the Americans. The United States was planning to test its MX intercontinental missile by firing it from California and letting it splash down in the Tasman Sea between Australia and New Zealand. Hawke agreed to help the Americans by letting aircraft engaged in monitoring the tests land at Australian bases. The news of this complicity outraged many in the Australian Labour Party caucus. It seemed that Hawke might not have the numbers to deliver on his agreement. Now he was on his way to Washington and, by some reports, was going to tell the Americans that he had changed his mind. Given the

American reaction to what they saw as my default, I wondered how they'd respond to Hawke. But they decided to offer a helping hand. George Shultz announced that Australian assistance wasn't really needed after all.

I took no comfort from the fact that Australia had been allowed to disappoint its powerful ally. It didn't seem that the Americans wanted to be disappointed twice.

7

Refusing to Back Down

INVECTIVE RAINED DOWN on us. Secretary of Defense Caspar Weinberger accused the New Zealand Government of making a serious attack on the ANZUS alliance. He told the Senate armed services committee that we were following a course which could be of great harm to us. A Republican from Maine who chaired the Senate sea powers and force projection subcommittee announced that he had asked President Reagan to employ punitive sanctions against New Zealand. He wanted quotas imposed on our American trade. If that didn't do the trick, he demanded that surplus American butter be dumped on world markets, to undermine our export business. A White House spokesman refused to rule out trade sanctions. A State Department spokesman said that the department was 'considering the implications for our overall co-operation with New Zealand' and that 'this consideration will be broad-ranging'. He, too, refused to rule out economic action.

Some of our exporters took all this very calmly. They probably calculated that the state of the market would decide what we sold rather than outrage at the Pentagon and the State Department. Others were anxious. Farmer organisations wrote to me demanding that the government indemnify them against losses caused by the nuclear-free policy. The tourist industry predicted an immediate fall in the number of North American visitors. Manufacturers of beer and ice cream were certain that their American customers would turn against them.

I had no reason to believe that the Reagan administration would itself impose economic sanctions on New Zealand. The information I had from Washington suggested that the outbursts from the White House and the State Department reflected the impulse of the moment and not any carefully considered intention. I relied on the assurance that George Shultz had given me that security issues and economic issues were separate. This assurance he publicly endorsed. Appearing before a Senate committee, he asked Congress not to pass legislation imposing economic sanctions on New Zealand. 'I don't think we want to transform an ally into an enemy,' he said. 'We believe that those who live by freedom and benefit by freedom ought to be willing to stand up and defend it, so we're disappointed in that aspect of the New Zealand performance, but basically New Zealand is a friendly country with similar values and we don't want to overreact to what they have done.'

What worried me more was American public opinion. Few Americans knew very much about New Zealand. Now we were all over the networks, handing out a snub to the United States Navy. This was dangerous. The row about the USS *Buchanan* would be a godsend to the American farm lobby if feeling against New Zealand was allowed to fester. When agricultural protectionism became patriotic instead of mercenary, we'd be in trouble. We'd be in more trouble still if American consumers thought we'd fallen into a crack between Albania and Nicaragua. We couldn't afford to have them engage in a personal economic boycott.

Before I'd heard the name *Buchanan* I'd planned a brief visit to the United States. Now I met a delegation of farmers and exporters and told them my strategy. Basically, I intended to get on American television as much as possible. I wanted to tell as many Americans as would listen that New Zealand's policy was directed not at the United States but at nuclear weapons. I wanted the American public to understand that New Zealand was a democratic country and that our decision to ban nuclear weapons was arrived at democratically. We weren't in there with Iran and Libya. On every issue except deterrence there were few differences between us. I told the delegation that I wanted to make it harder for the protectionist lobbies to target us by making us out to be an enemy. The delegation looked unconvinced. They agreed to wait to see what happened, but then they didn't have much choice about it.

Exporters weren't the only worry. Public opinion was far from overwhelmingly supportive of the government's handling of the ships issue. One poll showed that a majority (fifty-six per cent) were opposed to the visits of warships carrying nuclear arms. After that, it wasn't quite so clear cut. The same poll asked if its sample was in favour of, or against, the government's policy on nuclear warships. Forty-eight per cent supported the policy and forty-two didn't agree with it. I thought we could live with that. A greater potential difficulty was the continuing public attachment to the ANZUS alliance. Seventy-eight per cent of the people polled said that they were in favour of New Zealand's being a member of ANZUS. But our refusal to accept the visit of USS *Buchanan* meant that one way or another the alliance would be different in the future.

I wanted to hose down any feelings of insecurity before they got too firm a grip. One way to do that was simply to assert that nothing which had happened put the ANZUS treaty itself at risk. In point of law, that was correct. As a matter of fact, I had no reason at all to believe that either Australia or the United States would tear the treaty up. I tried to be as frank as possible about the likely effects of American displeasure on the operational side of the alliance, thinking that whatever happened was less likely to come as a shock if we were seen to have anticipated it. I knew that defence co-operation and intelligence sharing would be reduced, although I did not know exactly by how much. I was sure it would be a long time before any ship of the New Zealand navy was allowed to exercise with any part of the American fleet. I didn't see any great harm being done to us as a result. It was easy to look unworried about it.

It was also easy to wax indignant. In the early days of February 1985, when airwaves were full of accounts of the depths of American anger and threats of American retaliation, when headlines proclaimed the imminence of American trade sanctions, anger came naturally. The United States, I said, was blustering. 'New Zealand has got a government which has been straightforward with its people, where an election was fought with this issue being at the core of credibility, and where this government won a substantial majority and where this government enjoys very firm support from the electorate on that issue of principle. I regard it as unacceptable that another country should by threat or coercion try to change a policy which has been embraced by the New Zealand people.' I don't have any doubt that what I said struck a chord. Many people in New Zealand were less willing to accept American arguments that we'd put our security at risk when the arguments were made by a country that seemed to be running some kind of international protection racket.

Now we were embarked on a delicate balancing act. Chucking abuse at the Reagan administration served the political purpose of the moment, but in the longer term it would not help the government to keep faith with the electorate, to which it had promised not only a nuclear-free New Zealand but a nuclear-free New Zealand inside the ANZUS alliance. We certainly had no mandate from the electorate to spin off into non-alignment.

I was unhappy to see reports from the Soviet news agency Tass suggesting that we had done just that. Tass was writing us up as another milestone in the long march against American imperialism. I summoned the Russian ambassador to my office. Vladimir Bykov was the first of the new breed of Soviet diplomats to appear in Wellington; he was charming and polished and his suits fitted him. His pleasant wife was from the same mould as Raisa Gorbachev. I told Bykov of my annoyance that Soviet news reports were misrepresenting our position in a totally unacceptable manner. Look, I said to him, you just mind your own business. He nodded in what seemed to be understanding and agreement. But perestroika with Bykov didn't go any further than his suits. In the way of ambassadors, he must have kept on sending home reports which said that, thanks to his efforts, New Zealand was a cherry ripe for plucking, and Moscow must have taken him at his word. Many months later a senior Soviet official arrived here and announced his readiness to sign an agreement for naval co-operation between the Soviet Union and New Zealand. When I was asked about it at a press conference I said the only agreement I wanted with the Soviet Union was for the Russians to keep their ships as far away from us as possible. Soon afterwards, Bykov was recalled.

Those distractions aside, I wanted to make it clear to the public that it wasn't the New Zealand Government that had allowed our alliance with the Americans to become what George Shultz was now calling 'inoperable'. We weren't choosing to take a path to non-alignment. We were still a member of what I thought I might as well call the 'Western community', our membership of the Western alliance now being somewhat problematical. We were still on the side of the angels, even if the angels were strangely slow to recognise this. Wherever I could, I took the opportunity to point out that New Zealand was a starter for any defence of Western interests that didn't involve the deployment of nuclear weapons. Western interests, I argued, meant far more than being on one side in a nuclear contest. New Zealand was more than willing to pull its weight in any kind of conventional defence co-operation between Australia, New Zealand and the United States.

Armed with this, to me, compelling argument, I set off for Los Angeles. The first stop was Honolulu. Here, in accordance with time-honoured custom, I was welcomed by officers sent for the purpose by the Commander in Chief, Pacific. Custom also demanded that the hours spent waiting in the VIP lounge be passed in polite conversation. I could think of nothing we might talk about that would not lead those earnest naval officers into recollection of the rejection of USS *Buchanan*. Long minutes passed in agonising silence. At last we hit upon the *Cosby Show*. We talked about television programmes until dawn broke and the aircraft was finally readv to go.

In Los Angeles I was again the victim of the implacable courtesy of Americans. The State Department routinely offered Secret Service protection to visiting heads of government. For representatives of low-profile inoffensive nations like New Zealand, they provided a security detail and a bullet-proof limousine. More prominent, or controversial, figures got bomb dogs, a sharp-shooter defence against snipers and whole floors assigned to them at expensive hotels. I asked my official advisers to thank the State Department but tell them I preferred to be unescorted. Our consul general at Los Angeles cabled asking me to reconsider. New Zealand's actions had been widely reported in the news there recently, he said. There were many unstable characters around with easy access to a lot of firepower. Thinking that the consul would know what he was talking about, I took his advice and ended up going to Disneyland in the company of twenty-seven Secret Service agents.

In Disneyland I had a great time. The Secret Service men joined in the spirit of it. They were boyish in their coats and ties, each with what looked like a hearing aid in the ear and a lump like a hernia under the arm. Between the rides I met a family from home, who gave me the cheerful greeting all politicians expect on such occasions: 'We came here to get away from you!'

Next morning I rose early to do battle with the American news media. My first appearance was on morning television on the east coast. There was a makeshift studio in the lobby of the hotel. I could hear my two interviewers through an ear-piece, but there was no monitor to see them. This just about symbolised the level of understanding we reached. The two news presenters were in their way an echo of the State Department. Their tone was one of puzzlement. They seemed unable to understand why New Zealand didn't want to be defended by American nuclear weapons. I couldn't understand why they couldn't understand what seemed to me obvious. 'Look,' I finally said, 'there's only one thing worse than being incinerated by your enemy and that's being incinerated by your friends.' The voices in the ear-piece suggested that the screen, had there been one, would have been filled with blank looks.

This interview was typical of almost every television appearance I ever made in the United States. My interviewers were always impeccably courteous, even when expressing their disbelief. They were always bewildered by what to them was inconsistency between our refusal to be defended by nuclear weapons and our willingness to co-operate with the United States in other ways. If you wouldn't take nuclear weapons, you had to be missing a chromosome. They loved the line, an old favourite at the State Department, that New Zealand was along for a free ride, taking the benefits of American protection without paying some part of its cost. At that point I'd leap in and remind them that New Zealand soldiers sailed off to fight the Kaiser in 1914 and Hitler in 1939, adding that our commitment to democratic causes was such that we'd managed to fight one more war than

America had since 1945. It made little difference. Whatever we'd done in the past, we weren't on the team now.

My only appearance on American television I felt satisfied with afterwards, I made later in 1985 on Ted Koppel's *Nightline* programme. He was in Washington, I was in New York. Koppel was a tough interviewer, but he obviously wanted to draw out the argument and test it rather than hector me about it. He sent a crew to New Zealand to find out why a majority of its people endorsed the action of its government. His was the only programme I appeared on which was willing to acknowledge any perspective other than the view from the State Department. When I'd finished, something occurred that never happened to me before, and certainly hasn't since. The studio crew applauded.

It was always pleasant to find pockets of liberal opinion in the United States supporting New Zealand's stand against nuclear weapons. Many Americans took the trouble to write to me to tell me their feelings. Their letters far outnumbered the few I received from the States like the one addressed to the 'Commie President of New Zealand'. But I doubt in the end that New Zealand sank very deeply into the American public consciousness, if we sank at all, for good or ill. We were never a burr under the saddle like Iran or Nicaragua. We were simply too small, and the United States was too large.

I once went to a coastal city in California to receive an award from a local peace foundation in recognition of New Zealand's efforts towards disarmament. The members of the foundation were kind and intelligent, and warm in their welcome. About six hundred people, mostly faculty members and business people, gathered to see the presentation. The award itself was a handsome creation. Etched on thick glass was a map of the world, its centre the Pacific Ocean. Only one detail was missing. In the space to the east of Australia where New Zealand should have been, the map showed nothing but ocean. It's hard to make an impact when your country drops off the end of the world.

Before I left New Zealand I had asked for a meeting with the State Department. Here was a chance for the Americans to make a point. I was informed that an officer would meet me, but only to advise me formally of the administration's response to New

Zealand's refusal to receive USS *Buchanan*. The officer in question was William Brown, Deputy Assistant Secretary for East Asian and Pacific Affairs. He was not exactly the office boy, but in the world of diplomacy he might as well have been. I met him across the dining table at our consul general's residence in Los Angeles.

The atmosphere was tense. Bill Brown had not come to negotiate; he had come to deliver a message. In an even tone, he read out his list of the actions the United States proposed to take. There would be an end, forthwith, to all military co-operation. All postings of New Zealand military personnel to United States facilities were cancelled. We would no longer be invited to the confidential briefings the Pentagon regularly held for officers of America's military allies. The agreement under which the United States shared what was called processed intelligence with New Zealand was terminated. The United States would make intelligence material available to the other parties to the agreement only on the understanding that they did not pass the material on to New Zealand. That was it for defence and intelligence. To cap it off, there would be a diplomatic freeze. New Zealand's representatives were going to find it harder to get anyone in the administration to talk to us. For the record, I put New Zealand's position. We remained a member of the ANZUS alliance. We were willing to function as a member of the alliance to the best of our ability. We were now to be excluded, not by our choice but at the will of the United States. I told Brown that his list of reprisals would be seen for what it was, an overbearing response by a powerful partner to an ally that had forged through democratic process a legitimate policy. He murmured an acknowledgment.

One of the officials on my side of the table was Gerald Hensley, head of the Prime Minister's Department and bearer of disturbing messages from the intelligence community. Now his moustache seemed to quiver with outrage. Leaning forward, he argued forcibly that the intelligence relationship was built on mutuality of benefit. Whatever New Zealand got from the United States was surely matched by our contribution to American knowledge of the South Pacific. Again, Brown did no more than acknowledge Hensley's words. He didn't have a brief to argue with us.

Brown rose to leave. I led him to the door and down the driveway past the waiting representatives of the news media. He gave nothing away. It was then that I became aware that he had no car to pick him up; I walked with him to the gate and watched for a moment as he set off on his search for a cab. If he hadn't been under orders to give us the silent treatment, he could have had a lift.

The problem now was breaking the news to the New Zealand public. In less than two hours I was due to give a speech to a lunchtime audience in Los Angeles. Before I left for the speaking engagement we cobbled together a statement. Rather than downplay the impact of American actions, which might in the end incite the State Department to greater efforts, I decided to go for the moral high ground. I would acknowledge the seriousness of what had happened. I wanted to convey a sense of lofty regret that a great democracy had chosen to act against the interests of a smaller one. The statement said, in part, that what the United States had done was not 'the kind of action a great power should take against a small, loyal ally which has stood by it, through thick and thin, in peace and war'. Sometimes you have to over-egg the pudding.

Then I went off in my armour-plated limousine to deliver my speech at the Ambassador Hotel. My Secret Service detail was, as ever, in attendance. Dedicated to their calling, they could not contemplate anything so straightforward as going in the front door and through the lobby to the luncheon room. Instead, they took me to a goods lift, which creaked its way upwards and opened onto a kitchen — the room in which Robert F. Kennedy was murdered. Alone, as a tourist, I would have looked at it with respectful curiosity. In the company of twenty-seven Secret Service agents, I felt decidedly uneasy.

In the luncheon room itself, the guests sat at tables in front of batteries of cameras. I made my speech, adding to my text an account of my meeting with Deputy Assistant Secretary Brown. Again, I thought it best to appear pained by the American action. I told the audience that New Zealand's ability to play its part in the security of South-East Asia and the South Pacific would be diminished as a result of what had happened. 'I regret these moves,' I said. 'They are serious, and they will, to a degree, be damaging.' The audience looked sympathetic.

I'm sure that the State Department had no thought of our immediate capitulation. Its first object was to stand us in a corner as a sorry example for other allies. In the longer term, its choice of reprisals seemed carefully calculated to induce in the New Zealand electorate a profound sense of insecurity. In aiming squarely at the self-esteem of our military establishment, its strategists had introduced to a vital organ of the New Zealand Government a virus they hoped in the end would enfeeble us. They must have counted a lot on the treatment meted out to our military people. Their ostracism was almost complete. Our attachés in Washington, who played golf and pumped weights with their American counterparts, suddenly found themselves out in the cold. They were ostentatiously excluded from briefings and shunned on social occasions. No longer were our officers guaranteed places on training courses run by the American military. No longer did they enjoy plum postings at American defence establishments. Our naval vessels visiting American ports found themselves paying to use commercial bunkering facilities.

I don't think the Americans ever understood that all of this would be a matter of indifference to most members of the elected government. In a country like the United States where the defence establishment had enormous political and economic influence it must have been hard to come to grips with the lack of sympathy between New Zealand's government and its armed forces. I can't remember being in the least moved by the sufferings of our military representatives on the cocktail circuit. Nor, on the diplomatic level, was I ever much concerned by the frequent refusal of the Americans to speak to us. It just never occurred to me to measure our standing as a country by the yardstick of briefings at the Pentagon or invitations to the White House.

Among some of our defence establishment the nuances of ostracism were keenly felt. I have kept a memorandum written by Denis McLean, the Secretary of Defence, on the subject of a change of command ceremony at the headquarters of the US Pacific Fleet. The secretary and two senior military officers received invitations to attend. Although nobody was sure if the invitation was an error or an idiosyncratic gesture by the retiring commander, McLean decided to go.

Arrived safely in Honolulu, he found himself 'sitting under a large marquee with several hundred other guests, gazing at the bleak grey wall of USS *Antietam*'. Then followed the formal change of command. According to McLean, the speech of retiring Admiral James (Ace) Lyons was 'feisty'. New Zealand was ranked with Vietnam among American difficulties in the Pacific. As McLean described it, the admiral said that '"New Zealand's withdrawal from the fabric of ANZUS co-operation into what can only be termed 'essential neutralism' continues". (The Canadian Admiral sitting alongside me seemed to inch away a bit at that point.)'

Poor McLean then 'stood in line for a long time to shake the hands of Admirals Hayes and Jeremiah. Both seemed thunderstruck to be confronted by such an apparition as the New Zealand Secretary of Defence. I am told by the Australian High Commission that their post has reported that they were both very annoyed not to have known that an invitation had been issued — let alone accepted. Pacific Command's capacity for joint action and co-ordination in military matters is hopefully more sure than is indicated by this incident.'

McLean's sufferings as a wallflower told of more than the social setbacks our defence establishment had to suffer. It pointed to the deep resentment many officers of the United States Navy felt towards New Zealand after our refusal of the *Buchanan* visit. Among American naval staff the refusal was taken personally. Our diplomats in Washington always counted on a far more hostile reception at the Pentagon than they ever had from the State Department. Defense officials like Secretary Weinberger (the one who when asked why he hadn't invited New Zealand to a briefing said that he had 'lost the address') and Secretary of the Navy John Lehman were far more violent in their condemnation of New Zealand than ever was George Shultz and his array of under-secretaries. The great geographer Admiral William Crowe, when still Commander in Chief, Pacific, wrote to a senator in support of a resolution calling for sanctions against New Zealand, and appeared before the Senate armed services committee to testify that New Zealand's decision to ban American nuclear-capable ships was likely to increase the risk of nuclear war.

The following words of the admiral were perhaps most telling of naval sentiment: 'It is a little difficult for me to understand how the New Zealand Government feels the United States Government can obligate its men and women to the defence of a country that does not welcome those men and women and will not let its platforms come in.' I don't doubt for a moment the sincerity of Crowe's words. What he really couldn't understand was that we made a distinction between the men and women under his command and the weapons their vessels carried.

Another factor in American displeasure was the belief of their negotiators that the *Buchanan* visit would be accepted. Its eventual rejection was undoubtedly a shock to them. This is how, very shortly after the visit was rejected, Bill Rowling, New Zealand's ambassador in Washington, reported the views of Assistant Secretary of State Paul Wolfowitz.

Almost until the end, Wolfowitz 'had believed we were in sight of a favourable New Zealand Government announcement of a ship visit, which would test the US principle of "neither confirm nor deny" to the limit but would not have broken it. Quite unexpectedly that exercise had come unstuck.' After attributing the collapse of the proposal to a lack of political will, Wolfowitz said that he saw 'a fundamental contradiction in the New Zealand Government's attitude'. According to the assistant secretary, 'the precondition . . . of an accommodation was compromise and some initiative on New Zealand's part. The alternative was frank acknowledgment that ANZUS in its present form was no longer sustainable.' Not hearing the frank acknowledgement, the Americans assumed that compromise was on its way.

Rowling reported that he 'told Wolfowitz very frankly that I too had been taken aback by the way events had broken since my departure from New Zealand. I had thought . . . before leaving that a ship visit was something that my government could agree to . . . .' He added that, 'the Americans are shaken by the speed and apparent decisiveness with which a mutually satisfactory accommodation which was in the making, became unravelled.'

This is the reply that went back to Washington from Wellington: 'For a time it looked as though a way had been found . . . But it was not to be. Sparked off by some lamentable leaks and other manoeuvres, we had a movement of opinion here which no government could have stood against. You should make this clear to Wolfowitz. You should also make clear that the decision we took was in no way capricious or taken for doctrinaire reasons . . . The policy of no nuclear weapons on ships coming into our ports is about as firmly established as any policy can be. There is no alternative but to look ahead from that base.'

One of our troubles was that the Americans never seemed to hear the last part of that message. They were sure from the beginning that we would change our policy. From the nuances of negotiation they took only one meaning. As Wolfowitz suggested, our reluctance to abandon the ANZUS alliance was seen by them as willingness to pay for membership at the price set by the United States. I'm certain that they took New Zealand assertions about the integrity of our position as being only so much window-dressing, and worth about as much as the assertions of other governments which declared themselves to be nuclear-free and called upon their allies to 'respect the policy'. If they'd understood from the start that we really meant it, they wouldn't have been so disappointed.

Perhaps because they wanted to save face, some American negotiators took to claiming that they'd been misled. Some blamed New Zealand officials. That was unfair of them. Our officials had, after all, recommended that *Buchanan* be accepted. More often, it was me they pointed the finger at.

The most vehement of my accusers was H. Monroe Browne, the American ambassador. Not long before he left New Zealand, he unburdened himself of an assault in which he said that I had deceived American officials about USS *Buchanan*. The most persistently repeated of his claims was that when I met George

Shultz in Wellington immediately after the 1984 general election, I asked for six months' breathing space. This presumably was needed to get the Labour Party into a receptive frame of mind for a change of policy. It used to puzzle me. I knew that Browne was single-minded, but he never struck me as dishonest. It wasn't until I reread an account of the Shultz press conference, held the same day I met him in Wellington, that I worked out what might have happened. It was Shultz who said, before he saw me, that he understood that no ship visits were planned for six months. Browne can only have got press conference and meeting muddled.

Misunderstanding was later embroidered into assertion that had it not been for me, New Zealand and the United States might still be working allies. This I found increasingly tiresome. When I visited Europe in 1986 I said that I would resign as Prime Minister if the United States would simply accept without demur our nuclear-free policy. My offer went unaccepted. The fact of it was that personality was peripheral to it; it was the policy the United States would always find repugnant. However it might be described, in the evasions of diplomacy or the flourishes of politics, our exclusion of nuclear weapons was in itself a rebuke to the advocates of nuclear deterrence. The only way we could appease those advocates was to surrender to them, and in the end, we wouldn't.

8

The Oxford Debate

I WAS IN THE UNITED STATES on my way to England at the invitation of the Oxford Union. This long-established university debating society had a famous place among those who liked good public speaking and fierce intellectual contest. The Union liked to challenge what was ordinarily taken for granted and to shock conventional opinion into self-analysis by its choice of subject. It often invited guests from outside the university to speak about controversies with which they were associated. In 1984 the Union's secretary was a young New Zealand woman and it was at her suggestion that the Union asked me to take part in a debate on nuclear deterrence, in opposition to an American speaker, yet to be identified.

I was entranced. I was a debater at the University of Auckland, and Oxford was the ultimate in university debating. The Union debate would give me the chance to put New Zealand's case to a new audience uninfluenced by the distractions of New Zealand's domestic politics. In the heat of debate my argument would be refined and tested.

The Union asked me to argue that 'The Western nuclear alliance is morally indefensible'. This I refused to do. When it came to deterrence I thought that one side's nuclear weapons were as bad as the other's. I didn't want to be publicly critical of the West and not the East. At my insistence, the topic was changed to allow me to make the case that 'Nuclear weapons are morally indefensible'.

My official advisers counselled against my taking part. I would upset the United States. I would give offence to Mrs Thatcher's government. (In both those claims they were perfectly correct.) They saw nothing to be achieved by the debate. Nothing that might dissuade me escaped them. Early in 1985 the Oxford and Cambridge Unions held a joint debate. The *Daily Telegraph*'s account of the event was headlined 'Drunken uproar ends first Oxbridge debate'. The brawl that followed was blamed by Cambridge students on a large party of Oxford supporters. Four copies of this newspaper report appeared on my desk, sent by the diligent among my advisers.

The diplomatic establishment was at one in its concern with the British Government. The Ministry of Foreign Affairs reported from London that 'the British reaction to our policy on port access for nuclear warships has generally been excessive. Mrs Thatcher and her senior officials have expressed serious concern about the possible consequences and implications of our policy.' Like the Americans, the British said they were worried that our nuclear-free policy would have a ripple effect. Quite why they took it on themselves to turn New Zealand from its course I don't know, but they certainly did. The governments of Australia and the United States could, not unreasonably, claim that they had a right to remonstrate with their ally about policies affecting the alliance. The British Government had no such standing. We had the slenderest of military connections. The British used to argue that New Zealand's actions might fan public opposition to the deployment of Cruise missiles in the United Kingdom. Given that the Thatcher government seemed to rejoice in the embarrassment caused to its political opponents by the antics of protesters at Greenham Common and other missile sites, their worries about my capacity to rouse public opinion always struck me as especially hypocritical.

New Zealand's High Commissioner in London was on the point of retirement. His name was Bill Young. He was a National Party politician but was harmless enough. He once greeted a newly arrived Cabinet minister by saying, 'You wouldn't believe the price of vegetables in London!' Now he went to pay his farewell call on Mrs Thatcher. Part of the cable he sent back to Wellington read: 'Never in my life have I been subjected to such a firm barrage as the one I experienced . . . Mrs Thatcher opened up immediately on the matter of British warships entering New Zealand ports, saying that the current situation was little short of a tragedy. We have always marched together, she said, but in this case New Zealand Government policy meant the complete exclusion of British warships from our ports.'

I'm sure that the High Commissioner's account of the meeting was accurate. Mrs Thatcher always sounded in private exactly like she sounded in public. When I met her in London on my way to Oxford she spoke to me in slogans. She was a one-way communicator — I might as well have been sitting in the back row of a public meeting. She strode along, her handbag swinging in time with the points she was making; there was nothing personal, nothing appealing, in anything she said. It was all done for the record.

What she said was mostly a rerun of what she'd already said in public in the United States. There to tune up the special relationship, she insisted at a press conference that she wouldn't tell New Zealand whether or not British ships were carrying nuclear weapons. In response to a question from an American journalist she said, 'I am as disappointed as you are in the approach taken by the New Zealand Prime Minister.' I thought it pathetic that she should grovel to Ronald Reagan at the expense of a country that in its eagerness to help, and forgetting to allow for time differences, declared war on Hitler's Germany twelve hours before Great Britain did. What her harangue meant in effect was that our hospitality industry was to be deprived of whatever income it got from the infrequent recreational visits of Royal Navy vessels to New Zealand ports.

From time to time the British Government sent its agents to New Zealand to point out to us the error of our ways. The first of those was Baroness Young, a forthright woman. She was firm in her conviction that she was here on a mission to the heathen. The only notable incident of her visit was caused when a dish-eared reporter heard me say on her departure that she'd left her broomstick in my office. Another British envoy was Admiral Sir John Fieldhouse, easily the silliest visitor ever had. 'You,' he said to me, 'are inspiring communism in Africa.' When I opened the door to show the admiral out of my office, I saw to my astonishment a uniformed sailor, who stood rigidly, holding a plush cushion out in front of him. On the cushion rested the admiral's braid-encrusted hat. I couldn't resist. 'Not today, thank you,' I said as I swept past.

Above: *From this: a 1961 demonstration; Christians as the conscience of the nation . . .*
NZ Herald
Below: *. . . to this: the secularisation of protest. A recognition that there is nowhere to hide.*
NZ Herald

A question mark over the Pacific: the mushroom cloud rising from a 1973 French nuclear explosion, as seen from the decks of HMNZS Otago.
Ministery of Defence

With Secretary-General Javier Perez de Ceullar, the patron saint of the Rainbow Warrior *dispute, at United Nations Headquarters, 1984.*
UN Photo / Yutaka Nagata

American ambassador and racehorse owner, H. Monroe Browne.
NZ Herald

Outpost of empire? An aerial view of the village of Nukunono, the Tokelaus.
External Aid Division, Foreign Affairs

The seat of power, 1985. The office had recently been swept for 'bugs' following the leaking of details of a discussion with American Ambassador H. Monroe Browne, in an agency story from Hong Kong. The sofa was large enough to accommodate both of us. This is also where the CIA showed their home movies (see pp. 72–73).
NZ Herald

Above: *Hear no evil, speak no evil, see no evil. Australian Minister of Foreign Affairs Bill Hayden, New Zealand Minister of Foreign Affairs Warren Cooper, and United States Secretary of State George Shultz hold a press conference after the Anzus talks had ended in Wellington, July 1984.*
NZ Herald

Below: *Beware New Zealanders bringing flowers: with Vice-President George Bush at the White House, 1984.*
The White House

Bill Rowling at the time of his appointment as New Zealand ambassador to Washington, 1985.
NZ Herald

Irresistible force, immovable object. A handshake at Chequers, Mrs Thatcher's official country residence, 1984.
London Pictures Service

Before the hostilities commenced . . . Speaking at the Anzac commemoration, Rockefeller Centre, New York, 1989.
Wide World Photo Inc.

The British High Commissioner in Wellington (or as I once heard him describe himself, Her Britannic Majesty's representative in New Zealand) was Terence O'Leary, a man in the wrong line of work. Out of place in a modern foreign service, he should have been in the old colonial office. As far as I could tell, the people he cultivated in Wellington were drawn without exception from among the ranks of the critics of the government's defence policy. Whenever I planned a visit to England, he would come to see me to tell me what 'Margaret' would like to speak to me about. He was last heard of on the Wellington cocktail circuit complaining that he was the only British High Commissioner to New Zealand who had not been knighted. New Zealand under his stewardship had not abandoned its nuclear-free policy. 'Margaret,' the High Commissioner said, 'only pays by results.'

Now the High Commissioner came to see me bearing a Note from the British Government. A Note is a kind of low-level diplomatic protest. This is what it said: 'The Prime Minister and Ministers are looking forward to seeing Mr Lange early next month. They understand that he will also be speaking at the Oxford Union and proposing the motion "All nuclear weapons are immoral". As regards the debate, Mr Lange is of course perfectly at liberty to participate and speak his mind. However, British ministers are very concerned that by speaking in favour of the motion proposed he will be directly involving himself in a subject which is a major point of dispute between the government and the Opposition in the United Kingdom.' (The British Labour Party in those days favoured unilateral disarmament.) 'The press are likely to represent what is said as an attack on NATO, as detrimental to the Western position on arms control, and as adding to the difficulties already faced by some European countries.' (Here, I thought, the British press must be different from ours, which wouldn't represent anything as an attack on NATO unless you said to them, 'This is an attack on NATO.')

More followed. After reminding me that demonstrators had already begun to protest about a new Cruise missile deployment in Britain, the Note continued, 'British ministers are anxious to do whatever they can to ensure that the current difficulties within ANZUS are resolved as soon as possible and without acrimony. They consider that the sort of press publicity likely to be generated by Mr Lange proposing a motion on the lines apparently chosen will make this task a great deal harder. It will be virtually impossible for British ministers to avoid commenting on what Mr Lange has said, which means that Britain would be dragged in to the wider dispute.'

The authors of the Note ended by assuring me of their good intentions. 'The foregoing views arise solely from the concerns of British ministers about the damage which this action could do to British and alliance interests. No other party has been consulted. British ministers have also expressed the hope that if Mr Lange has not made a public announcement of his forthcoming visit and of the Oxford Union speech by the time this message reaches him then he will avoid doing so until he has reflected on the ministerial views given above.'

I told O'Leary not to worry. I had no intention of going to Britain to preach unilateral disarmament and I was not about to launch an attack on NATO's doctrines. The whole point about New Zealand's case was that it rested on an analysis of New Zealand's unique strategic situation, and that was the only case I wanted to make. I didn't like to say to him that the Note made me more determined than ever to go to Oxford.

Diplomatic uncertainty about my visit was compounded by public anxiety about New Zealand's trade with Europe. For most of its modern history, the greater part of the farm exports on which New Zealand depended had gone to Britain. The British trade produced the prosperity that marked my boyhood in Otahuhu. Then Britain joined what became the European Community, and our happy days as an outpost of British agriculture were over. To ease us over our difficulties, the British and the Europeans agreed that Britain would continue to import limited amounts of New Zealand farm products. In face of Europe's mountainous agricultural surpluses, these concessions were fiercely contested. At each renewal of them, their value was eroded.

To preserve what was still a major part of our export trade, New Zealand's diplomats worked busily in European capitals. In France and Belgium they pointed to the rows of white headstones marking the graves of New Zealand soldiers killed in the First World War. Throughout Western Europe, they reminded all who would listen of New Zealand's commitment to the Western alliance. New Zealand, they would say, could not carry its share of the burdens of the West if the West would not buy its lamb and butter.

Our disagreement with the United States showed the foolishness of basing an argument about trade on sentiment. The British were always willing to threaten us with damage to our European trade as a means of bringing us into line behind the nuclear deterrent. There were plenty of alarmists back home who were willing to believe the threats or use them as ammunition against the nuclear-free policy. The truth of it was that our wartime sacrifices and our support for Western interests didn't get our butter into Britain. It got there because people in Britain who bought butter liked New Zealand butter and always looked for the little anchor on the packet. It was British consumers who put pressure on the British Government to keep taking our agricultural products. As it was, I ground my teeth when I heard a pompous British minister remark that Britain was buying our butter because the British Government distinguished between its affection for the New Zealand people and the misguided policy of the New Zealand Government. One of the small and unexpected bonuses of the nuclear-free policy was its capacity to annoy the British.

The rest of the Europeans were absolutely hard-headed. Business was business. When I said in Europe that New Zealand was still aligned to the West, everyone believed me. They might have felt differently if I'd gone around the Continent inciting popular resentment against the Cruise missile deployments, but I didn't. Nobody on the Continent ever tried to read me a lecture on my errors of judgment the way the British did. Governments of places like the Netherlands and Belgium seemed worried about New Zealand only in the sense that we were irritating the Americans at a time when the Americans were already irritated enough by the growing public resistance to the Cruise missile deployments. Their concern didn't flow over into trade. When it came to our agricultural exports, what influenced the Europeans were the strength of their farm lobby and the size of the wine lake and the butter mountain.

When I arrived in London I found that my journey might have been wasted. The debate was nearly called off on the night before it took place. The Oxford Union told me that my opponent objected to the change of topic and wanted it altered to something closer to the original. I told them I would not debate anything other than the topic we had agreed on. My staff made tentative plans to have the High Commission invite an audience that I would address when I'd turned my debating notes into a speech. For hours, the debate hung in the balance.

My opponent was Jerry Falwell, evangelist of Lynchburg, Virginia. His participation was another reason why I was looking forward to going to Oxford. The founder of the Moral Majority represented everything I disliked about the corruption of religion by politics. Through this and similar movements, the church had been hijacked by right-wing politics, advancing an unlovely agenda under the guise of righteousness. To me, the doctrines of the religious fundamentalists seemed as dangerous and destabilising as the beliefs of any practising Marxist. Their attitude towards the possession and threatened use of nuclear weapons I found especially frightening. I could not dispute the view of many Christians that there were causes in whose service war was just. But the invention of nuclear weapons, with their capacity to destroy the planet, made mainstream Christians revise their thinking on just and unjust wars. Many reached the conclusion that the use or threatened use of such dangerous weapons could not be justified in any circumstances. Not so the fundamentalists. To them, deterrence was the only sure bulwark to be had against communism. At their most extreme, they argued that true believers would be snatched up to heaven at the moment of holocaust. Most disturbing of all, these unreasoning doctrines were echoed in the White House, where reigned the scourge of the Evil Empire. In debating Falwell, I wanted to challenge assumptions which were, temporarily I hoped, nesting in the heart of the United States Government.

Falwell in person, when I eventually met him not long before the debate took place, I found unattractive. He was wired for the contest as if we were about to take to the college football field. He couldn't or wouldn't make any effort at human contact. This confirmation of my expectations I found deeply reassuring.

My relief when I finally heard that the debate would actually take place was tempered by my growing nervousness. The *Buchanan* incident and the American reaction to it had put the nuclear-free policy into news reports all around the world. Television channels in Australia, Japan and the United States were going to report what happened at the Oxford Union. Most importantly to me, the debate was going to be shown live in New Zealand, where the time would be early on Saturday morning. I couldn't face going home again if I let New Zealand down. There wasn't any middle ground. If I didn't win the contest, the debate would be a disaster for the nuclear-free policy. The diplomatic establishment would do me if I fluffed it. The people who supported the policy would be grievously disappointed. If Falwell did me over, I would be handing deterrence a major moral victory.

I drove to Oxford with my old friend Joe Walding, our new High Commissioner, but not even his comforting presence made me feel less tense. Our arrival was disrupted by an anonymous caller who announced that he had placed a bomb in the Randolph Hotel where we were staying. The Thames Valley police arrived with the bomb dog. After a while, the Randolph Hotel was pronounced free of any peril greater than woodworm. None of this helped. I fidgeted through the afternoon.

Evening arrived and I strapped myself into my dinner suit. The Union building was jammed with people. On the way to supper, a Japanese television crew showed extraordinary self-confidence and unconcern for others by walking backwards in front of the official party up a winding staircase. The old debating chamber had a church hall feel to it. The audience filled the body of the room and the galleries above it.

The debate, when it finally started late in the evening, took the form of a series of set-piece speeches. On each side, as well as the two principal speakers, were student debaters and a British member of Parliament, Labour on my side, Conservative on Falwell's. Both the latter were engaging speakers. The Tory, I must say, won my heart when he began his speech by saying hello to his auntie in Onehunga, just up the road from where I lived. Speakers from each side followed one another in turn, the affirmative, my side, going first. Members of the audience were entitled at any time to interrupt and ask whoever was speaking to give way. The speaker might then, if he or she so chose, allow the interrupter to have the floor. This happened frequently. All the while the audience simmered and surged, now rippling with applause, now murmuring their agreement, sometimes hissing, sometimes jeering.

Once we got going I knew it would be all right. Oxford was a conservative university, and a victory for the affirmative side in the voting afterwards was by no means certain, but a large part of the audience obviously felt warmly towards New Zealand. They applauded vigorously before I'd said a word. The moment in the debate that stuck in most people's minds afterwards was entirely unscripted. I gave the floor to a young American, who was, I later learned, a Rhodes Scholar from the US Naval Academy. He certainly had a military bearing. He asked me, in view of New Zealand's decision to ban nuclear ships, how I could justify our membership of ANZUS. I said I'd give him the answer if he'd hold his breath a moment. 'I can smell the uranium on your breath,' I added. I heard later that they didn't like that at the Pentagon.

Falwell suffered more than I did from interruptions. He gave way to a young man who rewarded him by calling him the 'Reverend Ian Paisley of the sunbelt'. Falwell asked who Paisley was. 'The Ayatollah Khomeini of Northern Ireland,' came the reply.

Falwell spoke at length about the threat to civilised values posed by Soviet Russia. The fundamental aim of Russian communism was world domination and the West was justified in deploying nuclear weapons to deter Soviet aggression. He took up the point of other speakers on his side who argued that New Zealand was only allowed the freedom to disagree with its friends and allies because we were sheltering under the wing of the American deterrent. Falwell put it less pleasantly. He said, 'It is the moral code of the West to take care of those who cannot take care of themselves, or who don't wish to.'

For my part, I spoke about the essential irrationality of the arms race, which led to the greater and greater proliferation of weapons, even as the people who built them understood that they were daily adding to the risk of our total destruction. I got the warmest response when I referred to American efforts to make an example of New Zealand. We were being told by the United States that we could not decide for ourselves how to defend ourselves, but had to let others decide that for us. That, I said, was exactly the totalitarianism we were fighting against. The audience roared.

When the votes were counted, late in the night, the affirmative side won, 298 votes to 250. It mattered much more to me that I stood up for New Zealand and our nuclear-free policy and didn't fall over.

I went from England to Geneva, where the United Nations had a standing conference on disarmament. New Zealand was formally an observer at the conference, and I had been asked to speak to it. In its way, this was as great an opportunity as Oxford. My speech would not attract public interest the way Oxford did, but it would be read by most of those with any serious interest in disarmament. It was a chance to give substance to the argument for nuclear-free New Zealand and present our decision to preclude nuclear weapons as a reasoned and responsible means of arms control. I thought carefully about what I'd say. The truth of it was that I often started a speech with no idea in the world how I was going to finish it. Not at Geneva. The conference procedures required the text to be read and not departed from, so this time I wrote it all down. I went to the Art Deco Palais de Justice, where the League of Nations once deliberated, and the conference on disarmament now sat among the reminders of the disappointed hopes of another age. To the distinguished jury there, I put my case.

The starting-point of the argument for our nuclear-free policy was an analysis of the threat to security in the South Pacific. It was impossible on rational examination to see any serious threat there in the activities of the Soviet Union. The evidence was simply not convincing. In the South Pacific context, it made no sense to import a destabilising contest between the two great powers. New Zealand was not threatened by nuclear weapons, and it was worse than pointless to ask the United States to defend us with nuclear weapons. New Zealand's action in refusing a nuclear defence would not reduce by one the number of nuclear weapons in the world, but it showed that their deployment could be resisted when there was the political will to do it.

In the South Pacific, deterrence was a surrender to dangerous irrelevance. In practice, deterrence meant the inevitable escalation of nuclear armaments. The number of the weapons had expanded beyond the point of reason. Every refinement of strategy on one side was countered by a further refinement on the other. Every technical improvement was matched by another. As every move was met by countermove, the potential for destruction mounted.

I believed most existing measures of arms control to be a fraud. I wondered at the effrontery of the nuclear powers who offered an end to the arms race in the latest refinement in technology and trumpeted the weapons system which would for the moment give one the advantage over the other. It was the endless search for technological advantage that was actually driving the arms race. I despaired at arms limitation agreements that prevented their signatories from doing only what they had no intention of doing anyway. The partial test ban treaty, for example, outlawed testing in the atmosphere, above the atmosphere and in the sea. Its signatories did not want to test their weapons there. Underground testing went on unabated. A comprehensive test ban treaty never got off the ground. The non-proliferation treaty did nothing to prevent the nuclear powers from deploying their weapons wherever they could find countries willing to play host to them.

The only real restraints on nuclear weapons had come about when popular feeling had transformed itself into political will. Public unease drove nuclear testing underground. Even French nuclear testing was finally driven below the surface of Moruroa Atoll by concerted diplomatic action. Limits to the arms race were possible where the political will existed to impose those limits. In New Zealand there was will and opportunity.

Having placed the nuclear-free policy squarely in the context of South Pacific security, I had no easy prescription to offer the rest of the world. I didn't see any way that New Zealand's solution could simply be adopted elsewhere in the world. The fact of it was that the majority of people in New Zealand felt secure without seeing any need to be defended by nuclear weapons. In a country like the United Kingdom, the bulk of opinion seemed to go the other way (the British Labour Party went to some lengths to hide its policy of unilateral disarmament). Unlike the people of Western Europe, the people of New Zealand had no territory bordering on that of an alien, incomprehensible power. We didn't feel suddenly defenceless when the nuclear umbrella was taken away from us. Many people in Europe, on the other hand, were genuinely alarmed by the thought that the United States might pack up and go home. Knowing this, I was never an advocate of sweeping measures of unilateral disarmament. They might well cause more anxiety than the nuclear deterrent and so defeat their purpose. Instead, I saw it as the duty of responsible governments to look for means of securing the peace and safety of their populations in ways which did not involve an inevitable reliance on nuclear armaments. I told the conference on disarmament that 'We do not say to any country in the world, do as New Zealand does. All we say is that when the opportunity is given to any country to pursue a serious and balanced measure of arms control, then that country has a duty to all of us to undertake that measure.'

There were some in the nuclear-free movement in New Zealand and elsewhere who were disappointed that I did not simply urge the governments of other countries to follow New Zealand's example. While it was possible to be cynical about the willingness of other governments to look for real alternatives to a nuclear defence, I could hardly foist New Zealand's solution on them. The whole point of our policy was that it was right for our circumstances. It was our judgment, based on our assessment of what we needed for our security. Others had to work out their own way to disarmament. I couldn't do it for them, any more than they could tell me how New Zealand should be defended.

In its shorthand form, this approach came to be summed up in the saying that New Zealand's nuclear-free policy was 'not for export'. This catch-cry was, I know, widely used by our diplomats in their attempts to limit what they felt was the damage caused by the nuclear-free policy. In its abbreviated form, without its accompanying invitation to look for alternatives to deterrence, it became a nonsense. Our policy, in the end, was pointless unless other countries in their turn adopted genuine measures of arms control. What I hoped to do, not by offering answers for others but by describing what New Zealand had done, was to make the point that alternatives were possible. What we needed was the political will to look for them.

On my way back home from Europe I called at Singapore. Here, as the last remnant of the 'forward defence' practised by earlier New Zealand governments, was stationed a battalion of our infantry. (It was in fact one of our two battalions of infantry.) The soldiers trained in the jungle of Malaysia and lived, at immense cost to the taxpayer, on allowances which seemed to be paid with an eye to recruiting. Their presence in Singapore was an anachronism and planning was under way before the Labour Government was elected to return them to New Zealand. Now it seemed wise to leave them for a while. They were a sign to the rest of the world that we weren't neglecting our responsibilities to our friends and trading partners in the region and withdrawing into isolationism. To underline our commitment, I spent a hot sticky afternoon in the jungle training ground where soldiers dressed as trees popped up amongst the undergrowth.

This, it turned out, was a pointless exercise. Singapore put little value on the presence of our soldiers once we'd fallen out with the United States. Its government saw the battalion as a kind of trip-wire which, if disturbed, might lead to American involvement. Without the Americans, there was no possible use in one battalion. Prime Minister Lee Kuan Yew was very straightforward about it. He told me he liked the nuclear-free policy. It was the kind of romantic idealism everyone should go through when young. He himself, he said, had to be realistic. He lived at the crossroads of East and West. Why did I think he had a Soviet fishing base in Singapore? So he could have the US Seventh Fleet all around him and still be non-aligned. I'd learn, he said. Hoping to change the subject, I asked him about the future of the Philippines. 'What a terrible shame,' he said. 'All those years of dictatorship and they've still got a foreign debt!' He was from a hard school, Harry Lee.

Then it was home. At Auckland to greet me was a sight rarely seen — a demonstration in support of the government. Several hundred people brought banners and flowers to the airport to celebrate the nuclear-free policy. I saw in the company my brother. He who had never called me anything but David was carrying a sign that read 'Good on You, Dave'. It was a wonderful moment. Then I drove straight to a butter factory, where the New Zealand Co-operative Dairy Company, one of the country's largest exporters, had long before invited me to open an extension to their plant. Here I faced a huge gathering of farmers who weren't having any of this nuclear nonsense. It was threatening their business. They were polite but insistent. However promisingly they might begin, there weren't any easy days in politics.

9

Enter the *Rainbow Warrior*

THE ATOLLS OF THE SOUTH PACIFIC are formed on living coral, pushed up from the seabed to form a reef. On one side of the reef are the shallow waters of the lagoon. On the other there is nothing but a heartstopping plunge to the floor of the ocean miles below. In the shelter of the reef is an island, or more than one. Here grow coconut palms and other food and cash crops of the islanders. For much of the year there is an equable climate. There is no wealth, but there is always a sufficiency. The atolls are cocooned in distance. Many of them are very beautiful. Some of them are poisoned.

The pariah among South Pacific atolls is Moruroa. This island forms part of a territory called French Polynesia, which includes the famous islands of Tahiti and Bora Bora. France maintains that the islands are part of its metropolitan territory. At Moruroa, in keeping with the custom of other outside powers which use Pacific islands in their possession to store chemical weapons and dump toxic wastes, France tests nuclear weapons beneath the surface of the atoll. Objective scientific evidence of the effects of the testing on the atoll is hard to come by. France insists that the tests are safe, but then refuses to move the tests to European soil. It appears that severe damage has not yet been done to the atoll, but there is no evidence that rules out damage being done in the future. What is known is that any leakage of radiation from Moruroa would contaminate the migratory fish on which the people of the South Pacific depend, not merely for much of their income but for life itself.

In July 1985 the Greenpeace vessel *Rainbow Warrior* arrived in Auckland Harbour to prepare for its journey to the test zone at Moruroa. The Greenpeace aim, like that of the Labour Government when it sent a frigate to Moruroa, was to focus world attention on the atoll, helping to create a climate of opinion that one day might make further testing unsustainable. Greenpeace undoubtedly intended its vessel to make itself a nuisance to the French authorities, creating good television in the process. But on 10 July the *Rainbow Warrior* was torn apart by high explosives. One of its crew was trapped and drowned. The vessel itself was damaged beyond repair.

This was a time when feelings were running high in New Zealand about the rift in the ANZUS alliance. Occasionally I would get threatening letters from deranged individuals who had taken ANZUS to their hearts. My first appalled thought when I heard about the bombing was that it was an outbreak of indigenous political violence of a new and disturbing kind, but this surmise was soon dispelled. The police arrested a couple posing under the name of Turenge, who proved to be agents of the foreign operations section of the French secret service. They were part of a team of French agents sent to New Zealand to sabotage the *Rainbow Warrior*. Some of the others were identified by the New Zealand police, but France refused to give them up. Indeed, France refused for some time to admit any kind of responsibility for the bombing. Finally, the French Government was unable to contain the scandal. At the end of September the head of the secret service was sacked and the Minister of Defence resigned. Who in the French Government was ultimately responsible was never revealed.

Why France destroyed the *Rainbow Warrior* I do not know. This is the explanation given to a New Zealand MP by Michel Rocard, who became Prime Minister after the bombing. 'M. Rocard said the Americans and Soviets were some five or six years ahead of France in their nuclear weapons technology, except in one particular area, relating to the effect of nuclear explosions on the ozone layer. France had the lead in this area. In order to observe the effect it was necessary to be virtually on the site of the nuclear explosion in order to take vertical measurements and observations. M. Rocard saw no other explanation for the very costly and sophisticated equipment he said was on *Rainbow Warrior* — it was the result of US-Soviet co-operation to observe the French tests close up. This was not, of course, public knowledge, but it seemed to have been a matter of great importance for the US and Soviet Union to have such a highly technical spy system in place — why else had *Rainbow Warrior* been replaced within two weeks?' Rocard kept his face straight when he said it.

France's violent intrusion into Auckland Harbour was a plain violation of international law, a breach of the article of the United Nations Charter which binds its signatories to respect the political independence and territorial integrity of all other nations. On that article of the charter rests the moral authority of the Western alliance. Sadly, the leaders of the alliance seemed willing to set this point aside when it came to the bombing of the *Rainbow Warrior*. Eager as they were to condemn terrorism in its other forms, they were curiously silent about events in Auckland. The leaders of the West expressed not a moment's outrage about terrorism directed by a government against opponents of nuclear deterrence.

Such silence was disturbing in more than its hypocrisy. The courts of international law have no officers to enforce their judgments. There are no international bailiffs. International law relies for its effectiveness on the willingness of countries to abide by it. At most times, in most places, it serves the common interest to do so. At other times, international courts and tribunals may be seen as a useful means of settling a dispute, and countries will agree to bind themselves to the result. But when the law is breached it cannot be enforced, unless countries more powerful than the country in breach persuade the offender that failure to comply will result in its discomfort. The silence of the leaders of the Western alliance meant that there was little chance that pressure would be brought to bear on France.

The Turenge couple were soon identified as Major Alain Mafart and Captain Dominique Prieur. Late in July they were charged with murder, with conspiring to commit arson and with the wilful damage of the *Rainbow Warrior*. They were held awaiting trial at Mt Eden prison in Auckland. Rumours abounded that the French Government would send more agents to New Zealand to free them. I never saw any evidence suggesting that an attempt would be made, but it seemed unwise not to allow for anything the French might do. I did hear from the police that the agents themselves claimed to be in peril from their own people. Prieur in particular seemed genuinely terrified. Perhaps, as spies, they believed their own folklore. (They also used to complain about conditions at Mt Eden, alleging they amounted to psychological warfare. Having seen many clients waiting trial in the crowded old jail, I knew what they were talking about, but it wasn't deliberate.) The advice I had was that it would be wiser to move the agents into military custody. I didn't want them treated as prisoners of war instead of criminal defendants, so we got round the problem by designating the Papakura military lock-up a civilian prison and putting them in it. The army put up new barbed wire to secure it and guarded it day and night, but nothing happened.

Rather than tunnel under the wire of Papakura to free its agents, France adopted other tactics. By September it was rumoured that France was willing to resort to economic sanctions. Its most powerful card was the influence it exerted over New Zealand's sales of lamb and butter in the British market. As a leading member of the European Community, France could effectively kill the trade if it wanted. Now we heard it suggested that it had it in mind to do just that.

This threat emerged against a background of a rapidly deteriorating diplomatic relationship. I couldn't contain my outrage at what had happened and I didn't want to. The French Government divided its time between trying to cover up and making political capital out of the incident. President Mitterand flew to the test site at Moruroa. He described New Zealand as 'a country which serves as a platform and a relay for initiatives hostile to France', and said that the sending of the agents was lawful. The French called off a visit to Paris long planned by Deputy Prime Minister Geoffrey Palmer when Palmer announced that he intended to tell them exactly how unlawful French actions were.

I'd like to have kept them at a distance, but we needed to know what they were up to. Some kind of negotiation was the obvious answer. Instead of going to Paris, Palmer met the French Minister of External Relations in New York. They agreed that negotiations would take place, but the talks didn't last. On our side the aim was to see that those who had lost by French actions were compensated. On their side the aim was to get the agents back. Basically, they wanted to know our asking price. We said that the judicial process had to run its course. The talks collapsed.

The agents never stood trial for murder; they pleaded guilty to the lesser, but still serious, charge of manslaughter. I thought that a charge of manslaughter could be defended and that if the agents had given evidence of the planning of the operation and their involvement in it then the charge might be still further reduced. The result would be reflected in a lesser sentence. The shorter the sentence, the more likely would France be to accept the verdict. But the lawyer who represented Mafart and Prieur was under instructions not to defend the charge and they each got ten years.

It was now time to send a message to France that the agents were not for sale. I wanted the French to understand that heavy hints of economic sanctions would not work, and thought I could call their bluff. France had now acknowledged liability and it was still to me inconceivable that having done so, France would demand that its agents be immune from retribution. I was adamant in public that Mafart and Prieur would serve out their sentence in prison in New Zealand.

Talks between New Zealand and French officials started again. On both sides the aim was the same, as was the result. The talks spluttered out.

The two French agents did not appeal against sentence. Instead, France put on the economic screws. Our exports there met with unaccountable delays in customs. Bales of wool were torn apart in a search for heroin. The sale in France of New Zealand sheep brains was prohibited. Shipments of meat and seed potatoes to the French colony of New Caledonia were arbitrarily rejected. Those sanctions in themselves did not amount to much, but they were enough to convince me that our trade in lamb and butter was at grave risk. We took soundings among the other members of the European Community. They were not encouraging. I had to accept that it was the intention of France to deal a body blow to our farming industry. We were a victim of extortion and there was nothing I could do about it.

I sent officers of the Ministry of Foreign Affairs to meet the French with instructions to explore the limits of a settlement. They duly returned with the outline of an agreement. France would apologise. It would pay US$6.5 million in compensation. It would stop its interference with our export trade. In return the prisoners would be deported from New Zealand (by law, foreign nationals imprisoned here may be deported at any time before the sentence is completed). They would be held in custody on the French island of Mayotte.

Mayotte is an island in the Indian Ocean. I asked the New Zealand negotiators about it. A hell-hole was their word for it. I don't know if the French had fooled them, or if they wanted to fool themselves; subsequent enquiries showed it to be closer to paradise. This detail aside, I thought the proposal was about as much as we could hope for. The continuing custody of the agents was at least some acceptance by the French of the principles that had led to their imprisonment. Grudging acceptance and future good behaviour was about the limit of what we could expect from them.

This was in May 1986. Cabinet discussed the elements of the proposal. The consensus was, Mayotte apart, that it was workable. The task now was to invest an unsavoury piece of arm-twisting with the dignity of a properly arrived at international settlement.

In June I went to Europe, mostly to meet European Community leaders. The possibility of a settlement was not yet public. In Europe I learned that the negotiators had found a rabbit whose production from the hat would represent to the world a dramatic breakthrough in the diplomatic deadlock. This was the Prime Minister of the Netherlands, who, when I met him, duly proposed that France and New Zealand refer their differences to a third party for mediation. I announced his intervention at a press conference at the residence of the New Zealand ambassador in The Hague. The ambassador's wife had two alsatian dogs, which proceeded to chase a senior diplomat across the grass. 'They vill not harm him,' said the ambassador's wife, herself of central European origin. They bit the hapless diplomat. I sat describing the Netherlands' initiative to journalists and the alsatians came and slobbered over the microphones.

The name of the likely mediator was already agreed. He was Javier Perez de Cuellar, the Secretary-General of the United Nations. This was not yet made public when Cabinet met on 16 June to make its final decision on the agreement. I put the case for settlement; I put the case against. I invited the discussion of the Cabinet. One member spoke against a settlement. As a member of Cabinet in the last Labour Government, Fraser Colman was on board HMNZS *Otago* when it sailed to the test zone at Moruroa to mark New Zealand's protest. He was a quiet, mild-mannered man who thought it wrong in principle to pay the blackmailer. For the others, it was a matter of political calculation, the damage to our European trade set off against the political costs of releasing the agents to French custody. Basically the Cabinet wanted peace with a minimum of egg on the face. The real concern was with the distribution of the egg.

We agreed on the limits of the settlement. The amount of compensation would be left to the mediator. It was understood that France would lift its economic sanctions before the mediation process started, and would agree not to disrupt the trade in lamb and butter. For Mayotte was substituted the South Pacific island of Hao.

Hao was part of the group of atolls which contained Moruroa; on it was housed a forward support base for the testing programme. It was an isolated place, certainly not a tourist destination. Here Mafart and Prieur would live as serving officers for three years, not in conditions of imprisonment but in confinement and seclusion on the island. Finally we agreed that Perez de Ceullar should be appointed mediator.

This appointment was not immediately announced. Perez de Cuellar made it clear that he would not take the job if any matters of serious contention between New Zealand and France remained undecided, so we settled the terms, leaving only detail to the nominal mediator. His attitude reflected the reality of international relations. He had no power to enforce any settlement he made, and could rely only on the willingness of the parties to accept it. He could not afford from his own perspective to make a settlement that France would ignore, since his mediation would then appear ridiculous. He had to know before he started what was the acceptable result.

The charade played itself out and the appointment was made public. Tributes appeared in the news media to the wisdom of Perez de Cuellar and to his long experience in the settlement of international disputes. I expressed my confidence that his ruling would be 'equitable and principled'. Some news reports from New York suggested that New Zealand and France had agreed on the details of a settlement, and that Perez de Cuellar was effectively a rubber stamp. In Wellington the line was that there was no agreement, though both governments, through diplomatic discussions, had a knowledge of each other's position and understood the limits of the settlement. In this cloud of obfuscation we awaited the result of the mediation.

Perez de Cuellar made public his ruling on 7 July and it met with almost universal disapproval in New Zealand. The departure of the agents for Hao was seen by many people as a betrayal of principle. I can't say they were wrong but equally I couldn't ask our farming community to pay the price of being in the right. The bombing of the *Rainbow Warrior* and what followed it was a miserable episode. All it left me with was contempt for a country that thought its own self-importance gave it licence to trample over the rights of others. That in itself has been little comfort.

The last part of the Secretary-General's ruling provided for the settlement of any differences that might arise between New Zealand and France over the implementation of the agreement. If a dispute arose that could not be resolved through diplomatic channels, it would be referred to an international tribunal for arbitration. The tribunal would consist of one member appointed by New Zealand, one member appointed by France, and a chairman appointed jointly by both. The decisions of the tribunal would be binding on both countries.

This provision was soon needed. Mafart and Prieur took up their duties on Hao; Prieur was visited by her husband, but little was heard about Mafart. Then in December 1987, less than eighteen months into his three-year posting on the atoll, he was flown to Paris. He was said by the French to be suffering from intestinal problems requiring examination under general anaesthetic. Whatever the cause, his removal from the island was a breach of the settlement, which prohibited the agents' departure from Hao without the consent of the New Zealand Government. Our consent had not been asked for, let alone given. The French Prime Minister, Jacques Chirac, claimed that the agreement provided for an automatic return to France in the case of illness. He must have been thinking about something else.

Worse was to come. In May 1988 there was a presidential election in France. To boost his electoral prospects, Chirac's government negotiated the return of French hostages from Lebanon. A further token of the national self-esteem was still required. Prieur was now pregnant. Her father, from whom she had been many years estranged, was dying of cancer in Paris. That was enough. Home she went, over the protests of the New Zealand Government. Cameras and the Minister of Defence greeted her arrival at the airport.

Our first hope was that Chirac's defeat in the presidential election, and the appointment of Michel Rocard as Prime Minister, would put a stop to the posturing. Rocard, for all his commitment to the French nuclear deterrent and the Moruroa testing programme, was far more ready than his predecessor to acknowledge that France had acted wrongly in bombing the *Rainbow Warrior*. Before the election he told a visiting New Zealand minister that France had a terrible tradition of arrogance in its dealings in the Pacific. Less reassuringly, he thought it was unfair that Mafart and Prieur were punished when they were 'merely those who had been carrying out orders'. But Rocard did conclude by saying that 'ideally, and other things being equal, he would like to see his country keep its word'. After his appointment as Prime Minister he called the acting Prime Minister in Wellington and told him that it was shameful that France had gone back on its undertaking. Unfortunately this acknowledgment was as much as he could deliver.

We now heard from Rocard that his situation was very difficult. France was a chauvinistic country, the Prime Minister told us. There was little likelihood of getting Mafart and Prieur to return to Hao. Prieur was having a difficult pregnancy; Mafart was sick. More critically, the armed forces of France would not be able to understand or accept an order to return to the island. Against this military disapproval, the French Government could not stand.

He suggested that if New Zealand were to drop its claim for the return of the agents, France might offer compensation, 'something in the economic and trade spheres' was the way he put it. But we'd been once bitten. Cabinet agreed to seek only the return of the agents to Hao. I wrote to Rocard, 'The agreement which was reached between New Zealand and France to implement the UN Secretary-General's ruling caused me and my government great political difficulty. The fact is that New Zealand was outraged by the *Rainbow Warrior* incident. As to the two French officers who were brought before the New Zealand courts, I should point out that they were properly convicted of very serious criminal offences. New Zealand does not accept that military personnel acting under orders are exempt from personal responsibility for criminal acts. This is not a defence in New Zealand law, nor is it a defence in the legal systems of most countries. It is certainly not a defence in international law, as was clearly established in the judgments of the Nuremberg tribunal. Public opinion was overwhelmingly insistent that the two convicted officers should serve their sentences in New Zealand prisons. The UN settlement was therefore very unpopular. The fact that France has subsequently violated its undertakings under the settlement has only reinforced the public view that the settlement was inadequate.' I ended by expressing a hope for an honourable solution.

The French Government agreed that a doctor chosen by New Zealand should examine Mafart and Prieur to establish the exact nature of their medical difficulties. A New Zealand doctor living in London made the journey to Paris; he duly reported that he found Mafart in good general health. There was no swelling in his abdomen, and in fact no physical symptoms that might account for the intestinal pain Mafart reportedly felt. Our doctor concluded that future medical investigations appropriate to his condition could be carried out almost anywhere by a skilled operator, certainly in Tahiti and on Hao itself if need be.

Mafart's own doctor was concerned that his patient's psychological state might deteriorate towards depression, but agreed with his visitor that at present Mafart was best described as 'sad'. Sad though he was, Mafart was at least polite and co-operative. Prieur was surly. But from what the doctor could tell, the pregnancy was progressing normally. Our doctor was not left unmolested. His home in London was broken into. Papers were scattered. A radio was left playing on a French channel. On a desk was placed an ugly switchblade. The intruder was never brought to justice.

Diplomatic efforts to secure the return to Hao of the agents were, unsurprisingly, fruitless. New Zealand invoked the arbitration provisions of the Secretary-General's settlement and a tribunal that would hear our case against France was appointed. New Zealand's representative was Sir Kenneth Keith, a distinguished jurist. France chose Jean-Denis Bredin, a lawyer who counted among his clients the disgraced Minister of Defence, Charles Hernu. The chairman of the tribunal was Dr Eduardo Jiminez de Arachaga, a Venezuelan of standing among practitioners of international law.

It was agreed that New Zealand's case should be put to the tribunal by Geoffrey Palmer, who, as well as being Deputy Prime Minister, was Attorney-General. In 1989, when the hearing was yet to take place, Palmer succeeded me as Prime Minister. I in turn became Attorney-General, an appointment I held outside the Cabinet. It became my responsibility to lead the New Zealand case. I soon heard of dismay in the Ministry of Foreign Affairs, which told the new Prime Minister that my involvement would make our conduct of the case too 'political'. Before this was reported to me, I was largely indifferent to the arbitration. Now I assumed that the ministry's wish not to be provocative towards the French was a sure sign that the diplomatic establishment was planning on surrendering. I looked at the task with real enthusiasm.

I buried myself in the detail of the brief. I wanted to put New Zealand's case to the tribunal as forcefully as I could. I always knew the limits of what I could achieve. Even if it agreed with New Zealand about the nature and seriousness of France's breach of the settlement, the tribunal was unlikely to make an order for the return of the agents to Hao. No more than Perez de Cuellar could the tribunal enforce a judgment unfavourable to France against a country that plainly had no intention of abiding by such an outcome. The best I hoped for was a strongly worded condemnation of France's actions.

The tribunal's hearings were conducted in October 1989 in a suite at the Waldorf Astoria Hotel in New York. Our taxpayers will be pleased to note that our share of the substantial costs of this were met from the interest accumulated on part of the compensation money paid by the French in 1986. The New Zealand delegation took up residence in a more modest establishment. The work of preparation we carried out in the abandoned offices of our consul general in New York. This post was closed, but its floor-space in the Rockefeller Centre had not yet been given up. We lived on food delivered to the offices from the boundless resources of New York, and on a supply of Australian meat pies which I found left behind in a freezer and heated in a microwave. In the largest of the rooms we installed Professor Bowett, a greatly distinguished English scholar of international law, whose work with us gave much of the substance to New Zealand's argument.

In the suite of the Waldorf Astoria sat the tribunal on its pedestal, its clerks in front of it and, chain-smoking in their little booths, the interpreters. The proceedings were conducted in an air of assumed civility. New Zealand appeared first, uneventfully. Essentially we relied on the nature of the agreement between France and New Zealand, and its standing in light of the established principles of international law. We asked the tribunal to seek France's compliance with the terms of the agreement, and return the agents to Hao. Then passed a day in which France refined its response. The French presentation on the third day was characterised by a considerable degree of contest between the large number of people they had brought. The opening statement was marked by increasing agitation among the advocates who were to follow the leading counsel, and, as its length unfurled, by signs of downright hostility from a woman at the back. This curious spectacle was viewed impassively by two representatives of the French military, who sat like hovering angels of death at the end of the suite.

The French argued that any default on their part could be remedied in ways short of the return of the agents. In any case, they said, the passage of time was now against a remedy. It was best that the whole matter be forgotten, in the interests of good relations between France and New Zealand. They also argued that their action in removing the agents from Hao was justifiable, if not by the terms of the agreement, then on humanitarian grounds. The sufferings of Mafart and Prieur were rehearsed at length before the tribunal. It was, the French protested, incomprehensible to them that New Zealand was so far removed from the normal patterns of civilised conduct that it could not understand the moral imperatives of France's actions. There was a lot more in this vein.

After another day's interval, both countries made their final submissions. I wanted to appeal to the tribunal's sense of fairness. I spoke of the trust New Zealand had always placed in the forums of international law. I said that the law was deficient if it invariably served the interests of the strong and neglected the interests of the weak. It was wrong if the whole body of the law was corrupted by some easy avoidance of it by a country that had the power to resist its obligations to others.

The French representative reminded the tribunal of the realities of politics. He stated that his government would be unable to comply with an order for restitution. The French government might agree to return the agents to Hao atoll, but the French military would refuse to carry out its orders. What he was saying was that France was above its treaty obligations. It had entered into them but now did not wish to comply with them. That was all there was to it.

The tribunal retired for some months to consider its judgment. The house in Wellington of New Zealand's representative Kenneth Keith was broken into. All that was taken was the word processor on which he had been drafting his contribution to the tribunal's proceedings. A carving knife replaced it.

The tribunal's judgment was finally made public in May 1990. It did not, of course, order the return to Hao of the two French agents. In the end it had no option but to bow to the fact that it couldn't. But it was unanimous in its disapprobation of France's conduct, and strong in its language. It was as much as I hoped for.

Those ultimately responsible for the sinking of the *Rainbow Warrior* are still unpunished, and the testing at Moruroa still continues. Nothing in French history suggests that it will stop until countries more powerful even than France put a stop to it. I can only look forward to the day when they will want to.

10

An Ally Becomes a Friend

THE DECISION TO TURN AWAY USS *Buchanan* was hardly made when the man from the State Department threw out a hint that all was not yet lost. Or at least, he threw out a hint of a hint, enough to be detected by the trained mind of a diplomat. The State Department's representative was Paul Wolfowitz, Assistant Secretary for East Asian and Pacific Affairs. He was perhaps the least uncompromising of the American negotiators, which is not to say he was accommodating. He always liked to talk; he was the soft cop. He met our ambassador in Washington to tell him of the department's disappointment at the collapse of the *Buchanan* talks. In the midst of his recital of our failings, some mysterious diplomatic nuance surfaced. The ambassador reported to Wellington, 'I detected from the remarks a disposition on his part, after a decent interval and the venting of some US spleen, to respond to any move from our side to pick up the pieces of this aspect of the relationship.' This aspect of the relationship was ship visits.

Wellington responded with the hope that the United States might agree to send an FFG-7, one of a class of vessel that was not capable of carrying nuclear weapons. 'After a time we need an "easy" visit or two . . . and some fresh thinking.'

Nobody in Washington or Wellington was able to detect even the faintest signal that the United States was willing to send an FFG-7, or any other inoffensive vessel. I still thought it best to keep in touch with the Americans; I wanted time. The suspension of American military co-operation effectively put an end to the ANZUS alliance, and a lot of people in New Zealand weren't ready to hear that yet. They needed to believe that Big Brother was watching out for them. I had to keep arguing that the treaty still existed, and I didn't want the Americans saying otherwise. Until the country got used to the idea that the Americans weren't there anymore, I had to keep talking to them. But the focus of the talks shifted.

In its 1984 election platform, the Labour Party promised that New Zealand's nuclear-free status would be more than a matter of policy; it would be written into law. Some members of the party's caucus wanted to legislate immediately on the basis of the Nuclear Free New Zealand Bill, the bill that gave the former Prime Minister his reason for calling the general election. I didn't see it as a high priority. No law could for the moment do more than we had already done simply by announcing that nuclear-armed and -powered vessels would not be given permission to enter New Zealand ports. The caucus understood, but the nuclear-free movement outside Parliament didn't. Its leaders were always ready to see any delay in the legislation as a surrender to American pressure. This rain-dancing only made me less inclined to bring a government bill into Parliament but *Buchanan* changed my mind. One of the Cabinet suggested when we were discussing the *Buchanan* visit that we might as well legislate. It was a timely point. Given the howls of outrage when the Americans stubbed their toes on the bottom line of our policy, we might as well post a warning that the bottom line was still in place.

In one sense, New Zealand's capacity to legislate for its nuclear-free status was limited. Any country can, of course, make law that runs throughout its land area. It was easy enough to legislate to prevent the storage of nuclear weapons in New Zealand, or the dumping of nuclear waste, or the deployment of nuclear weapons in New Zealand territory. But international law put limits on our ability to control the passage of ships and aircraft. We couldn't prevent a nuclear-armed vessel sailing close to our shores if the vessel was exercising what was called its 'right of innocent passage' through our territorial waters. What we could do was restrict the entry of foreign vessels to what the law called our 'internal waters', in other words, our ports and harbours. It was the government's intention to write this narrow restriction into law which came to preoccupy the thinking of negotiators from the United States and New Zealand.

Broadly speaking, the law had two purposes. Much as I resented the accusations of untrustworthiness flung at me by the leaders of the nuclear-free movement outside Parliament, I accepted that the law would have standing as a yardstick of government conduct. The government could not break the law or allow it to be broken. If members of the public believed that the law had been broken, they had recourse to the courts. I was never certain how far the courts would go in reviewing a government decision to permit or refuse a visit by an allied naval vessel. Traditionally they were reluctant to put security issues to the test. More recently their reluctance was less obvious. The Americans expressed outrage at the thought that intelligence assessments of naval armaments might find their way into court, but judicial review was never a possibility that could be entirely excluded. The law's other purpose was more bluntly political. Once passed, it stayed in force until repealed. Any future National Government would be bound by it until repealing legislation had gone through the parliamentary process. The National Party was reluctant to concede in public that its fondness for ANZUS amounted to support for nuclear deterrence. Repealing the nuclear-free law would force it to declare itself.

The United States made it plain from the start that it saw the legislation as locking the door on ANZUS and throwing away the key, a violation of its policy of neither confirming nor denying the presence of nuclear weapons on board its ships and aircraft. If its ships came to New Zealand in compliance with the law, then, said the Americans, the ambiguity that surrounded them was immediately dispelled. We argued this point. We did not propose to put legal responsibility on the commander of any visiting ship. The burden of responsibility would lie on New Zealand authorities. It was the Prime Minister who would be bound by the law to refuse entry to any vessel, unless satisfied that the vessel was not carrying nuclear weapons. We would not ask the Americans to assist us in making our judgment. We would make it from our own resources, and they could say as much or as little as they liked about the status of their vessels. It was at the meeting of the South Pacific Forum in Rarotonga in the Cook Islands in 1985 that Bob Hawke said to me, 'If you don't ask them to tell you, you'll be right. They'll probably buy it.'

They never did. The fact of it was that the Americans were

never willing to send an unambiguously nuclear-free vessel to New Zealand, and they saw the law as shutting down the last chance that New Zealand might adopt the blind-eye policy of other governments. But we plodded on, negotiators on each side hoping, I suppose, that the simple fact of talking might one day drag the other into complicity. What our side was offering were refinements in the wording of the bill which would make it plain that the decision to admit a vessel or refuse admittance was New Zealand's. One example of this is probably enough. Early versions of the critical clause in the legislation required the Prime Minister, in making the decision about the entry of a vessel, to obtain reports from the Chief of Defence Staff and the Intelligence Council as to whether or not it was likely that the ship was carrying nuclear weapons. Amazingly enough, the United States took exception to those references as an unnecessary telegraphing of the American position. It really didn't make any difference, so out they went. We never did get around to deciding how many angels could dance on the head of a pin.

The bill was supposed to make its first appearance in Parliament in May 1985. Its introduction was postponed to allow time for discussion of the draft with the Americans. The most eager of our negotiators was the Deputy Prime Minister, Geoffrey Palmer. Palmer as a young man had studied law in the United States. Before he went into Parliament he was at an early age a professor of English and constitutional law. He loved America. He loved its Constitution. Any constitution was a thing of beauty to him, but the American Constitution he found particularly appealing. He believed in the tolerance of American society. In moments of emotion he used to describe the United States as the inheritors of Jefferson. He believed sincerely that a settlement of the difference between New Zealand and the United States was possible. He did not for a moment wish to surrender the nuclear-free policy. That policy was arrived at properly and democratically, which in Palmer's mind was enough to invest it with a certain sanctity. He went to the United States knowing that his task was to prepare the Americans to accept what they didn't want to accept. He believed they would, because he was going to appeal to their democratic instinct.

Palmer planned to visit the United States in October 1985. A few weeks before he left, Cabinet reviewed our negotiating position. It was formally agreed, and recorded in the minutes, that the government's objectives were to preserve intact both its nuclear-free policy *and* a beneficial working relationship with the United States. It was also agreed that the Deputy Prime Minister would engage in discussions with administration officials in Washington on the basis of paragraphs four and five of a memorandum that went to Cabinet over my signature and under the title 'ANZUS/Ship Visits'.

In this memorandum was the distilled experience and opinion of our diplomatic establishment. It began by setting out the possibilities ahead of us. We might be able to maintain the existing stand-off (in other words, the alliance wasn't operational, but the Americans contented themselves with the military sanctions they had already imposed). Next, we could have what the memorandum called 'a rupture of the security relationship with the United States'. Finally, we could make a serious effort to come to a working relationship with the Americans.

The authors of the memorandum viewed the second of these possibilities very seriously. This was their advice to the Cabinet. 'A rupture with the United States would do serious damage to our national interests. This might involve not only the American renunciation of their security guarantees to New Zealand under ANZUS, but also have adverse spillover effects on our important wider political and economic/trade relationship with the United States. Our relations with Australia, the United Kingdom, Western Europe, Japan and the South Pacific would also be affected.'

What would bring about this rupture was the passing of legislation that constituted an absolute ban on port visits by nuclear vessels. 'Frankly,' said the memorandum, 'it would be a waste of time for the Deputy Prime Minister to go to Washington to discuss such a proposition.'

At the same time, the legislation could not merely be an unenforceable declaration of New Zealand's nuclear-free status. The compromise, which paragraph five of the memorandum suggested, rested on the Deputy Prime Minister's convincing the

Americans that their 'neither confirm nor deny' policy would not be subverted if New Zealand alone made the judgment about the weaponry of visiting vessels. Paragraph four argued that an understanding with the Americans depended on more than agreement about the wording of the legislation. We needed ship visits. As the memorandum put it, we needed 'resumption of *some* ship visits, consistent with our anti-nuclear policies. This might begin in the first half of next year with a USN FFG-7 vessel known to be nuclear-free, to be followed later in 1986 by a Royal Navy ship. Thereafter further visits will take place in accordance with the government's non-nuclear policy.' How they read that in Washington, I can't tell. In the Cabinet in Wellington, it seemed fair enough.

Off Palmer went to Washington. There, in the State Department, he met George Shultz. It might have been one of the shortest meetings on record between a Secretary of State and a senior member of a friendly foreign government. The way I heard it, it lasted twelve minutes.

Palmer began by announcing that the purpose of his visit was to resolve the differences, perhaps the most serious ever, that had recently arisen in the relationship between New Zealand and the United States. He acknowledged that the Americans had objections to the impending legislation. But he wanted to make it clear that the legislation did not amount to a prohibition on ship visits. It set up a process by which decisions about visits might be made. Some decisions in favour of visits would doubtless be made. Shultz wondered what might happen when a ship arrived that was carrying nuclear weapons and an unfavourable decision was made. To mount a challenge to the doctrine of 'neither confirm nor deny' was not, said Shultz, the correct response to the growing threat from the Soviet Union.

Palmer must have seen an opening here. He said in his sincerity that New Zealand's policy was an imperative arrived at by means that were the antithesis of the denial of freedom of choice characteristic of communism. Carrying as it did the implication that the attitudes of those who did not accept this point were less than creditable, this sally was unfortunate.

Palmer's own report of the meeting referred to 'the chilly correctness of Mr Shultz'. Umbrage was certainly taken. Palmer's report continued. 'Despite the very considerable efforts we made to show reasonableness and flexibility, within the clear framework of our anti-nuclear policy, Mr Shultz, to the evident dismay of some of his senior officials, cut us off with the proposition that there was no way through unless we changed the fundamental thrust of our policy; that we cannot and will not do.' Shultz ended with some words of consolation, or possibly warning. 'Shultz indicated that while the introduction of our nuclear-free New Zealand legislation would provoke some American review of ANZUS, he did wish to see the relationship between our two countries remain one of friendship.' That said, Palmer found himself out on the pavement.

Palmer was hurt by the reception he got from Shultz. Pressed by the news media for comment immediately after their meeting, he expressed what I'm sure was real disappointment that the Americans hadn't met the New Zealand Government's proposals with what he called matching flexibility. The Americans were duly outraged at what State Department sources told journalists was a 'partial and inaccurate' characterisation of their position. The same sources said that Palmer had presented Secretary Shultz with a 'take it or leave it' option and that Shultz had decided to leave it.

Some American officials were more welcoming to Palmer. As he put it in his report, 'I was not able to resolve the ships visit question, but in the longer term it may well turn out that the visit to Washington did have a positive outcome.' What was needed in the meantime, Palmer concluded, was 'a quiet period, in which we keep ANZUS and ship visits off the headlines'. This, he believed, 'will enable those forces which favour a stable relationship between New Zealand and the United States to work'.

Exactly what those forces were, I wasn't sure. The merit of keeping ANZUS out of the headlines was no more clear. What was clear was that the passage of the legislation would provoke the United States into holding the funeral rites for the ANZUS alliance. It was time to start saying that the corpse wasn't worth shedding tears over.

I went to Christchurch to speak to a Labour Party meeting,

where in front of me sat party members, faces etched with suspicion. The Palmer negotiations were to them the prelude to a sell-out. But they liked the speech. I told them there was no going backwards, that the defence relationship between New Zealand and the United States could never return to its former stature. Then I got stuck into ANZUS. The treaty was not a formal guarantee that the United States would come to the defence of New Zealand. It was a commitment to consult, no more than that. The United States saw the ANZUS alliance as part of its global strategy of nuclear deterrence. It had resisted well-meaning attempts by New Zealand to use the ANZUS treaty as a framework for low-level military co-operation. Here was the crunch. If ANZUS was merely a nuclear alliance, there was no point in New Zealand being in it.

This speech was both preceded and followed by a ritual to which I became accustomed. My morning in Christchurch was punctuated by anxious phone calls from John Henderson, who had replaced Gerald Hensley as head of the Prime Minister's Office. In his earnest way, John reported that the Ministry of Foreign Affairs, having read my speech, was sure that I was signalling that I was ready to take New Zealand out of ANZUS, and blow sky-high their hopes of an accommodation. I told him to tell them that it wasn't over yet. After the speech, John's phone rang hot with complaints from the American embassy, the general burden of which was that the Ministry of Foreign Affairs had lost me off the leash. A senior official of the ministry was once heard to greet the American ambassador on the morning after one of my more heartfelt efforts with the words 'I didn't write it!' I loved it.

Not long after Palmer came back from Washington I said goodbye to an old adversary: H. Monroe Browne had finished his term as American ambassador in Wellington. At our last, informal, meeting I took him to task for his extraordinary assertion that I asked George Shultz in July 1984 for a breathing space of six months.

'Well, I thought that's what you said,' said Monroe Browne.

'If you thought that you obviously weren't hearing it,' I replied.

'Pardon?' said the ambassador.

We talked about horse racing the rest of the time.

Browne's successor as ambassador was a professional diplomat, Paul Cleveland. He didn't take up his appointment until early in 1986. He was a harder character than Browne but was easier to deal with. There was nothing idiosyncratic about him. He never gave me the impression he was fighting for a cause; he was just doing his job. He was always straightforward. 'That will cause you a lot of trouble,' was his usual line in warnings, though I sometimes heard of him letting down his professional guard at cocktail parties. He was at a reception given to mark the visit of some Canadian naval units when he launched into the Canadians for coming to New Zealand and shattering the solidarity of the Western alliance. But sailors are used to that kind of language.

Our nuclear-free legislation was still awaiting its debut in Parliament when I went to New York at the end of October 1985 to speak to the General Assembly on the fortieth anniversary of the United Nations. The place was full of heads of government and foreign ministers. President Reagan gave a reception for the lot of us in the glittering ballroom of the Waldorf Astoria. Under the chandeliers the tables sagged with food, and behind them were gathered ranks of journalists. On the floor of the ballroom the outline of a pair of shoes was marked, and painted in front of them the words 'Mr President'. On this marker the President duly stood. We all lined up to shake his hand. Here I had a demonstration of Reagan's remarkable evenhandedness. We were in order of the alphabet, Netherlands, New Zealand, Nicaragua. The Netherlands was a trusted ally. New Zealand was a known delinquent. Nicaragua had long since been carted out with the diplomatic rubbish. But each of us got the same greeting, and each in the same tone of voice. 'Hello,' said the President. I didn't know whether to be anxious or reassured.

Assistant Secretary of State Paul Wolfowitz was at the President's party, and wanted to talk. He told me that I shouldn't interpret George Shultz's remarks to the Deputy Prime Minister as being either final or unduly negative. Wolfowitz at any rate didn't see the negotiating process as being played out. If we kept in touch with the Americans about the legislation, it might not necessarily be a breaking point. So we had another crack at it. Back in Wellington, I announced that a senior official of the Ministry of Foreign Affairs would go to the United States for new talks about the nuclear-free bill. Then we got the word, very publicly, that the official wasn't welcome in Washington. I don't think Wolfowitz was trying to set us up. I think it was simply that his colleagues at the State Department had come to feel that New Zealand officials were no longer representing the views of the government.

The diplomatic freeze was getting frostier. New Zealand ministers found it impossible to get in to see anyone who counted in Washington. In the wake of the Geneva summit meeting between President Reagan and General Secretary Gorbachev, the Americans sent high-ranking officials round Asia and the Pacific to brief governments on what had happened. New Zealand was ostentatiously left off the itinerary. The irony was too obvious to pass without comment. 'I think the spirit of Geneva lasted about twenty-four hours when it comes to an ally,' I said at a press conference. 'I hope it lasts longer with their friends in Russia.' I announced at the same time that the nuclear-free bill was about to be introduced to Parliament. There wasn't much more to say to the Americans. 'It is their requirement that we do not proceed with it,' was how I summed it up to the representatives of the news media.

Before the bill was introduced, it went for scrutiny to the Labour Party caucus, where for two days members pored over its every word. The inevitable misgivings of the party outside Parliament were largely contained when the bill received the endorsement of a member of Parliament called Helen Clark. In those days she was a pillar of the left, and had many close connections in the wider nuclear-free movement. After characteristically thorough examination she reached the conclusion that the bill was the most practical mix possible and gave it what amounted to a seal of approval. The party followed her. Approved by the caucus, the bill received its first reading in Parliament.

Our last attempt at negotiation was made after the bill appeared in public. By now we were running short of people the

Americans would talk to. Our last card was David Caygill, the presentable and inoffensive Minister of Trade and Industry. It was agreed that he would meet a representative of the State Department under conditions of great secrecy at the house of our consul general in Los Angeles. The secrecy was at the insistence of the State Department. Having told the world that they weren't interested in talking about the legislation, they didn't want to be overheard doing just that.

The department's representative was James R. Lilley, a deputy assistant secretary. The meeting began with Caygill passing over to Lilley the 'non-paper ' with which he had come prepared. The non-paper was no imaginary document. It was, in diplomatic parlance, a document whose existence its authors would disavow if it ever came back to haunt them. It bound them to nothing. Caygill's non-paper was the memorandum that, several months before, the Cabinet had adopted as the basis of Palmer's talks with Shultz. Lilley, meantime, had come prepared with a non-paper of his own. It was, to say the least, to the point.

It began with the usual warning: 'You should understand that enactment of ship access legislation which reinforces the existing disruption of our co-operation as allies will prompt us to redefine our treaty obligations to New Zealand's defence and could result in the termination of our alliance relationship. The government of New Zealand should recognise that the issue is not simply the defence relationship but the whole fabric of special privileges and influence that New Zealand has enjoyed as an ally.'

There followed a critical analysis of the legislation, prefaced by the observation that the fact that comments were being made should not be taken as an indication that the United States would find legislation acceptable. Lilley's non-paper ended with a statement of what the Americans wanted.

'We request the following responses from you:

'1. The government of New Zealand's commitment to restore normal port access during the life of the current government, and not to rule out any type of US ship.

'2. Indication when the government of New Zealand would be prepared to receive the first US warship under neither confirm nor deny, other than an FFG-7.

'3. Indication when the nuclear-powered warship ban will be removed.'

Caygill's report to me concluded like this: 'Clearly a substantial United States concern is the impact of what happens in New Zealand on their interests world-wide . . . They gave two illustrations. First, they mentioned that Norway, Denmark and Japan had all expressed interest in the specific language of our legislation and had sought to know whether it had American approval. More importantly, both Lilley and Glassman indicated there had been strong Australian pressure on the United States not to cut a deal with New Zealand.'

On that note ended our attempts to reawaken in the Americans the spirit of Jefferson.

Australian Foreign Minister Bill Hayden and Secretary of State George Shultz planned to meet in San Francisco in August 1986 for the talks that had replaced the annual meeting of the ANZUS council. It was understood by our foreign service that the two would use their meeting to put a formal end to America's treaty obligations to New Zealand. Then we heard that Shultz and I would both be in Manila in July. I didn't see much point in talking to him. On the other hand, it seemed churlish to avoid him, so we set up a meeting.

What brought us to Manila was the annual meeting of ASEAN, the Association of South-East Asian Nations. After their meeting, the ASEAN countries consulted with the representatives of other countries with an interest in the region. The talks in 1986 took place amidst an air of celebration. Cory Aquino was only recently elected President of the Philippines. George Shultz, and Mrs Shultz, sang to us all at the closing banquet (so, unfortunately, did some rowdy Australians). At our meeting in my hotel suite, Shultz was conventionally businesslike.

We talked for a while. Shultz was matter-of-fact. This time I didn't get any sense of a contest. I thought that, in his own mind, he had already settled New Zealand and wanted to get on to more important business. I asked him why he couldn't accept that our legislation didn't amount to a request to the United States to breach its policy of neither confirming nor denying the presence of nuclear weapons. I told him we had no intention of making public our rejection of the visit of any particular vessel. He said that we might be able to keep it a secret, but the United States couldn't. 'Our system leaks like a sieve,' he said.

Then I asked him why he couldn't make an exception for New Zealand. I gave the example of Denmark. No nuclear-powered vessel of the United States fleet had visited Denmark since 1963. If a concession like that was possible, I didn't see why another couldn't be.

This must have tried his patience.

'You've got to understand,' he said, 'if you want ship visits that satisfy us, you've got to accept that from time to time there will be nuclear weapons on United States ships.'

Such are the absurdities of deterrence that this was the only time I ever heard a senior official of the United States acknowledge what every schoolchild knew to be the fact. Shultz ever afterwards denied having said it.

Our meeting lasted about fifty minutes. Shultz finished by confirming that he planned to take the final steps to lock us out of ANZUS when he met Bill Hayden in San Francisco. We agreed that our discussion should stay between the two of us.

We went out into the corridor. I heard a low roar. Shultz's secret service men thundered towards us, two alsatian dogs lolloping in front of them. Behind them pressed a crush of journalists. Thump! Back against the wall I went. From this new perspective I heard Shultz, to my astonishment, tell the reporters the outcome of the meeting. New Zealand was defaulting on its military obligations, he maintained, and the United States did not see why it should continue to extend a security guarantee. 'ANZUS is there and we have no plans to alter it at all,' he added. 'We will continue to operate under the treaty. The problem is it's minus one member for the present.' Some moments later he was telling reporters that the government of New Zealand would not accept the policy of neither confirm nor deny. That was too much. I interrupted him to say that we certainly were not challenging neither confirm nor deny. Our policy was not to have nuclear weapons in New Zealand.

Shultz wasn't happy. Oddly enough, I was. I don't think I

ever felt more certain than I did after that meeting that our policy was the right one. I didn't care about the withdrawal of a meaningless security guarantee; it only meant our freedom from the sham of a nuclear defence. After Manila, I could only see us going forwards.

At my last encounter with Shultz, a challenge to neither confirm nor deny was mounted more brutal than any New Zealand had ever uttered. It came at the press conference which marked the end of our meeting with the ASEAN ministers. A line-up of foreign ministers, among them myself and Shultz, faced the news media. In rows right in front of us sat our official advisers, including several representatives of the State Department, and behind them, the journalists. Nobody among the questioners was much interested in ASEAN. They wanted to know about nuclear-free zones. They wanted to know the future of American bases in the Philippines. They wanted to know if the bases played host to American nuclear weapons. In the chair was Vice-President Laurel of the Philippines. He was asked if the Americans brought nuclear weapons to their base at Subic Bay.

'No.'

'How do you know?'

'Because,' said Laurel, 'the United States would tell us if they did.'

From the front row, an American voice uttered, 'Jesus Christ!' I turned to look at Shultz. His face was totally expressionless. I think he earned whatever the American public paid him to represent them.

Back in Wellington, I found the Cabinet unperturbed by events in Manila. We were now increasingly confident that our trade with the United States would not be affected by the disagreement about defence. Irony in fact abounded. The Australians, loyal allies of the United States, now learnt that a valuable export market was threatened by American subsidy of wheat sales to the Soviet Union. So incensed were the Australians that threats were made to boycott the San Francisco talks. Trading in the United States was no easier for New Zealand than it ever was. But it wasn't made noticeably more difficult by the row over ANZUS.

In the end, the Australians turned up in San Francisco. Shultz and Hayden jointly announced that New Zealand's policy detracted from individual and collective capacity to resist armed attack. They added that, 'The United States said that it could not be expected under these circumstances to carry out its security obligations to New Zealand. Accordingly, the United States side stated that it is suspending its security obligations to New Zealand under the ANZUS treaty pending adequate corrective measures.'

New Zealand was no longer an ally of the United States. From now on, the Americans formally classified us as a 'friend'. Hayden, thinking of the wheat, asked if Australia could be treated as a friend and not an ally. But then he had the comfort of the nuclear deterrent.

11

The Point of No Return

WHAT HAPPENED IN MANILA would have seemed like a nightmare when I was Leader of the Opposition and worried about the strength of public attachment to the ANZUS alliance. On that score there was less cause for concern in 1986. It was the government's good fortune that events in Manila and San Francisco took place against a backdrop of steadily growing public support for the nuclear-free policy.

We had come a long way in a few years. At the end of the 1970s a survey showed that over sixty per cent of the public were in favour of visits by United States ships carrying nuclear weapons. Fewer people were in favour of visits by nuclear-powered vessels (a point which again showed the strongly environmental origins of the nuclear-free movement of the 1970s), but here, too, those who were in favour were a majority of the sample. The poll taken not long after the *Buchanan* incident which showed that fifty-six per cent of the population now opposed visits by warships carrying nuclear arms came as a great relief to me.

Some part of the shift in attitude was doubtless a result of the activities of the organised nuclear-free movement. It was now widely known by the name of the peace movement. In the early 1980s there were more than three hundred recognised peace groups. One of their goals was to have local authorities declare their territory to be a nuclear-free zone. In this, they were largely successful. By the time of the 1984 general election, ninety-four local bodies had declared themselves nuclear-free, and more than half of the country's population lived in self-proclaimed nuclear-free zones. Sceptics found it easy to sneer at the essential impracticality of the zones, but their educative effect was great. New Zealand local authorities are not by and large hot-beds of radicalism. When so many of them declared themselves to be nuclear-free, they bestowed on the policy a new respectability.

The nuclear-free zone I always liked the best was that declared by the Borough of Devonport. In 1983 I went there to have morning tea with the mayor, in celebration of the tenth anniversary of the departure of a Royal New Zealand Navy frigate for the French nuclear testing site at Moruroa. This determinedly pacifist borough was the home of New Zealand's principal naval base.

I was not directly involved in the peace movement in the 1980s. After the Labour Party became the government in 1984, I found its leadership more and more difficult to deal with. This was partly the result of the inevitable distancing between any protest movement, naturally impatient for action and intolerant of impediment, and any government, characteristically inclined to make haste slowly. The leaders of the movement chipped away at the government, ever anxious to find some dilution of the purity of our nuclear-free convictions. Its leaders made it plain that they wanted to see the nuclear-free policy cemented in legislation, not merely to tie the hands of some future National Government, but because they did not trust the existing Labour Government. This did not make me warm towards them.

My disenchantment was also the result of an alteration in the aims of the peace movement. Its great goal of the 1970s had been reached — nuclear weapons had been banished from New Zealand. Its leadership now had to find new targets. The peace movement became a home for a grab-bag of causes. You couldn't rise to prominence in it simply by proclaiming your opposition to deterrence; you had also to declare your commitment to land rights for the Maori people, freedom for gays and an end to the logging of native trees. It was hard to deal with its leaders when I couldn't be sure exactly what kind of offence was going to be taken next.

As the 1980s passed, the movement became steadily more radical in its objectives. Some of its members would have us disavow any form of defence force apart from what was needed to clean up after natural disasters. I was often at odds with them over the government's intelligence-gathering activities. Their favoured target was an installation called Tangimoana. This was a radio listening post, which lent its uninvited ear to signals sent by people without the sophisticated technology to guard against intrusion. I went there once to see what they did and how they did it. When the inspection was over I asked the supervising officer if I could listen to the news on the national broadcasting network. He hadn't counted on this. He produced an old radio which he made to work by rigging a wire to a radiator pipe to get a signal.

This officer was thought by the peace movement to be instrumental in the targeting of American nuclear missiles. A leading activist once picked out an old Post Office microwave dish and said that it was on site to pick up satellite signals. So it might have been, if the satellite was sitting in a nearby cowshed. A great deal of information about Tangimoana, and the post that replaced it, eventually found its way into the public domain. If the peace movement had protested about its true purpose, I might have found it harder to be unsympathetic. As it was, I was only irritated by their readiness to cry wolf.

Its commitment to radical causes effectively marginalised the peace movement in the late 1980s. Having helped put the nuclear-free movement into the mainstream, peace activists took themselves out of it. I found it useful to slam them publicly whenever they said anything particularly wacky; it was a way of demonstrating to the public at large that although the government might be quarrelling with the United States, we hadn't surrendered to the flakies. The irony of this hasn't escaped me. When I was young, the National Government cried communist at the sight of me and my fellow protest marchers. In a sense, the nuclear-free movement had gone full circle in thirty years.

What helped keep nuclear-free sentiment in the mainstream was the mainstream political process. The Labour Party was not alone in its support for the cause. Four political parties fought the 1984 general election; only the National Party did not promise the electorate that it would ban nuclear-armed and -powered vessels from New Zealand. At the end of the four-way contest, sixty-three per cent of the popular vote went to the three parties with nuclear-free policies. The figure didn't mean much in itself, since the nuclear issue would have been a deciding factor among only a few voters. But the fact that nuclear-free policies were adopted across a broad spectrum of political opinion gave them a standing greater than the Labour Party by itself could give to them.

The range of political opinion in the 1984 election was indeed extraordinary. There was, although I'm not quite sure at what point on the spectrum to place it, the Social Credit Political League. The league was the direct descendant of the disciples of Major Douglas, who wanted to fight the Great Depression with unlimited free credit. In New Zealand, the Douglasites were absorbed by the Labour Government in its first term of office, then spat out by it. Since then the league had carried on a lonely battle on behalf of strange economic theories. It had many supporters in rural New Zealand, enough to elect a member of Parliament at irregular intervals. In the early 1980s, when the National Government of Robert Muldoon was unpopular and the Labour Party was struggling to present itself as a credible alternative, Social Credit gained a lot of ground among urban voters.

Its original adherents were honest toilers, poorly dealt with by the divine lottery, folk who were ready to believe that their adversities were the result of international conspiracies. Zionism was regarded with much suspicion. The World Bank and the International Monetary Fund were thought to be plots against the innocent. Social Credit's distrust of ANZUS flowed, I suppose, from its instinctive dislike of entanglement with a large power of capitalist persuasion. At any rate, its two members of Parliament voted against the National Government and in support of the Nuclear Free New Zealand Bill in 1984.

The New Zealand Party was the creation of one individual, property investor Robert Jones. Jones was a scourge of the last Labour Government and then a fierce opponent of Robert Muldoon's National Party. By his own account, he despaired of Labour's ability to defeat Muldoon unaided, and so set out to peel away a slice of National's vote by inventing his own political party. His success was almost immediate. He didn't, of course, calculate on becoming the government. From the security of substantial self-esteem he said what he thought without fear of electoral consequence. The electorate loved it. Jones preached the economics of the self-sufficient individual; he also decided that his would be the first peace party. His defence policy, as I read it, was to have no defence policy. Subjected at one public meeting to the persistent interruptions of a heckler who doubted his credentials as a pacifist, Jones declared that he was about to come down off the stage and knock the heckler's bloody head off.

Jones was always armoured by his fine sense of the ridiculous. He was once defamed by an American publication which claimed that he, Roger Douglas and Ron Trotter were agents of the KGB. Jones was a self-made millionaire of impeccable capitalist credentials. Roger Douglas was Minister of Finance in the new Labour Government. (He leaned so far to the right that he became an advocate of the flat rate of taxation, which, according to David Stockman, the Reagan administration rejected as too unsettling for the voters.) Ron Trotter was an industrialist, the model of a captain of industry. Jones sent me a copy of the letter he wrote, urging legal action on the other two, stressing how much they might recover and lauding the wonderful American device by which your lawyer only charged you if your case was successful. There was a postscript. 'Of course,' he wrote, 'this all falls down if Ron Trotter really is a KGB agent.'

The New Zealand Party argued that money spent on defence was money wasted. Worse, it was fraudulent. If the country didn't spend another dollar on the military, it would be no more at risk than it was before. There have been times when I've thought that too, but in 1984 this radical stance didn't really match the temper of the electorate. Not everyone shared Jones's confidence that all who really wanted to could stand up for themselves.

Those who seemed most in need of assurance were the former leading lights of the military establishment. The leaders of the Returned Servicemen's Association were always ready to pronounce the government's quarrel with the United States to be a slight on the sacrifices of the past. Every Anzac Day, retired senior officers would try to take advantage of the emotion of the moment by charging the government with having betrayed New Zealand's allies. One former air force chief saw more than foolishness in the government's actions. He charged one Anzac Day that I was a captive of the Kremlin. I wrote him a letter, thanking him for all he had done to make the government's position more acceptable to the public. I added that his allegations about the Kremlin were doing more for me than for him, and could he please repeat them just before the Labour Party conference. His reply suggested he wasn't happy about being taken lightly.

This was not my only military correspondent. During 1985 a number of retired senior officers wrote to me to argue that the government should abandon its nuclear-free policy and pave the way for New Zealand's return to an operational ANZUS alliance. Among them were four former chiefs of defence staff and twelve former chiefs of staff of the three services. Their private correspondence proving ineffective, they decided to carry the battle to the public. They issued a statement, which was extensively covered by the news media, and at the core of which was the following charge: 'It appears to us that the heart of the current problem is the government's belief that the South Pacific region should be regarded as uniquely privileged to stand apart and yet secure in a divided world; secondly, that the nuclear attitudes of others have no relevance to New Zealand's interests or security; and thirdly that a modest contribution of effort towards meeting minor threats to stability within the region is an adequate provision for overall defence.'

From these premises, the retired chiefs argued that it was the policy of nuclear deterrence which had effectively prevented war for the last forty years and was the only sure means of doing so in the future. 'This,' said the officers, 'is a reality which all members of the public should recognise.' The officers were critical of the government's attempts to keep the country nuclear-free while remaining in ANZUS. 'As has been demonstrated in recent days, New Zealand simply cannot have it both ways. The two standpoints are irreconcilable.' None of this would have happened, they suggested, if the public were better informed of the value of nuclear deterrence and able, as they put it, 'to reach more logical conclusions'.

This body of unreconstructed military neanderthals spoke in the voice of a class apart. I gave them a status they didn't deserve when, asked by a journalist for comment, I said that I wasn't going to engage them in public, and added that 'these geriatric generals can carry on like this for as long as they like'. This unfortunately memorable alliteration led to a great parliamentary moment. Frank O'Flynn, Minister of Defence, was at sixty-six the oldest member of the Cabinet. In Parliament the day after the retired chiefs made their statement, a member of the National Party asked him if he was able to confirm to the House that a number of the so-called geriatric generals were recently retired and distinctly younger than the minister himself. O'Flynn replied stiffly that he regarded part of the question as somewhat personally offensive. But he had come prepared; he read out a list of the names and ages of every officer who had signed the statement, finally reaching Brigadier J. T. Burrows, aged eighty-one. Said O'Flynn, 'He taught me at school!'

This distraction aside, the former chiefs of staff were inadvertently helpful to the government. New Zealand, after all, was not a country in which arguments in favour of nuclear deterrence were routinely tossed around in public. Singing the praises of deterrence was more likely to frighten than to comfort. It was the generals' attempt to educate people in reality as they saw it which helped cast the debate into a choice between a nuclear future inside ANZUS and a nuclear-free future without a defence relationship with the United States. That was a contest the advocates of deterrence could not win.

In saying, as they did, that acceptance of visits by nuclear ships 'represents no more than very modest dues owed to an alliance partner on whom we rely to underwrite our ultimate security', the generals got their public relations hopelessly wrong. People could still tell a protection racket when they saw one. It was also characteristic of opponents of the nuclear-free policy to lash out at the simplicity that was an essential part of its appeal. They did their cause no service by suggesting to the general public that people would appreciate deterrence if only they could understand it. It's fatal in politics to put it to the voters that they don't know what's going on. You might as well tell them straight out that they're stupid.

In the tactical mistakes of its opponents, the New Zealand Government was enormously lucky. I used to complain long and loud when American ambassadors or visiting members of Congress lectured the public on our failings as an alliance partner. This, I protested, was interference in New Zealand's domestic politics. The truth of it was that I loved it. Every American intervention got up the nose of someone in New Zealand; Washington's ultimatums provoked more resentment that they did compliance. The United States made an even more basic tactical error when it openly favoured National Party politicians over members of the government. Washington doors slammed shut in the faces of ministers opened wide for Opposition politicians. In the 1987 election it was easy for the government to present the National Party as Washington's creature. You don't often get that kind of advantage handed to you on a silver plate.

What also encouraged nuclear-free sentiment was the simple fact that it was the policy of the elected government. The contrast with Australia makes the point. Both Australia and New Zealand were countries in which by the early 1980s there was considerable support for the nuclear-free movement, although it was far from a majority. Both countries were allied to the United States. Both countries were equally at risk from Soviet aggression (in other words, they weren't). It always seemed to me that Australia was nearly as likely as New Zealand to go nuclear-free. The Australian Labour Government decided that it wouldn't, and flung itself instead into the arms of deterrence. Whatever might be thought of the aims of its policy, its execution was highly successful and nuclear-free feeling in Australia was left to simmer. In New Zealand, the Labour Government set itself up as an active opponent of deterrence. What had been establishment thinking about defence was challenged not by street protesters but by the government. Cabinet ministers were heard to suggest that the old way of working under ANZUS wasn't all it was cracked up to be. Simply by lending the authority of government to the concept of a nuclear-free defence, the New Zealand administration stoked up nuclear-free sentiment.

None of this meant that I could ever take public opinion for granted. I seized upon the opinion polling when it appeared, and by the end of 1985 I was sure that a comprehensive inquiry into public attitudes was needed. As a result of the collapse of military co-operation between New Zealand and the United States, the government had decided to recast its defence spending. This exercise took the form of a defence review. Defence reviews had been held before, but had always been the exclusive property of the government's official advisers. These officials assessed the threat (in the days of the National Party's forward defence policy the threat was the toppling of dominoes in South-East Asia), and advised the government as to the appropriate military posture to adopt in response. I was unconvinced of the ability of our military advisers to give the government advice entirely undiluted by their desire to be friends once again with the Americans, but even more strongly did I believe that the incestuous defence reviews of the past took no account of an essential element of security policy. They did not ask what people needed to make them feel secure. The most sophisticated system of defence seemed to me not worth the money spent on it if it didn't reassure the people it was supposed to be protecting. Unlike the former chiefs of staff who thought it their duty to educate the public in military reality, I believed that military thinking would greatly benefit from the common sense of the general public.

To gather public opinion, and measure it as far as it could be measured, the government set up a body which went by the name of the Defence Committee of Enquiry. Its terms of reference were carefully defined. The government had published a paper called *The Defence Question* which set out a range of possible options for the organisation of the country's defence. The members of the committee were to invite public submissions on this paper and were then to question those who made submissions. They were to commission polling in order to provide objective data on public attitudes to defence and security. Finally, the members of the committee of enquiry were to examine the mass of information they would undoubtedly collect and sum it all up in a report which would then be fed into the defence review.

Much thought went into the composition of the committee. I didn't want people who would give inevitable endorsement to the government's position. I was determined to include at least one who would be seen as unsympathetic to the government's policy. For the latter purpose, a retired chief of military staff was the most likely candidate. This plan nearly came unstuck when the retired chiefs publicly declared themselves to be not merely unsympathetic to, but active opponents of, the government. In the absence of any better alternative, I had to persist. The military representative on the enquiry was Major General Brian Poananga, a former chief of general staff. To balance the military perspective, we appointed a prominent member of the peace movement, Kevin Clements. To reduce the imbalance of the sexes, we appointed Diane Hunt, a scientist of whose opinions I knew nothing. My hope was that this unlikely mixture of individuals would either stand back and allow public opinion to speak for itself, or, if they felt duty bound to sermonise, would speak in such contrasting voices as to effectively cancel one another out.

The enquiry was chaired by Frank Corner, a former Secretary of Foreign Affairs whose career in diplomacy had been a long one; as a young man he had taken part in the negotiation of the ANZUS treaty. That should have been enough of a warning. But I thought he had the standing to satisfy the public that the government's interest in their opinions was serious and its methods scrupulous. And I did not for a moment believe that he would use the enquiry as a platform on which to stand while he belaboured the government.

What went on inside the committee of enquiry, I don't know. Submissions certainly flowed into it, and it held a series of public hearings. An opinion poll was designed and at no little expense carried out. But the commentary written to accompany the committee's findings gravely disappointed me; the hope of a contest of ideas was not realised. The committee spoke in the voice of a retired diplomat. There was a hymn to the ANZUS alliance and the guarantee of security that it ostensibly provided. The committee talked about visits by allied ships that 'might' be carrying nuclear weapons, when, as all the world knew, some of the ships that once came here were certainly nuclear-armed. It talked about 'negotiating mistakes' which had contributed to the breakdown of the relationship between New Zealand and the United

States. The committee had apparently decided to prepare for the future by looking determinedly backwards.

Worst of all, it seemed to me that the committee had manipulated the poll data to produce one finding, duly sensationalised in the news media. According to the committee, a majority of the public were willing to see a resumption of nuclear ship visits and pay this price for an operational ANZUS alliance. The committee's opinion polling discovered that thirty-seven per cent of its sample wanted ANZUS with nuclear ship visits, forty-four per cent wanted ANZUS without nuclear ship visits, and sixteen per cent wanted out of ANZUS. The forty-four per cent who supported ANZUS without nuclear ship visits were then asked, 'If staying in ANZUS without nuclear ship visits proves impossible, what is your next choice?' Their preferences were then distributed between the supporters of nuclear ship visits and the opponents of the alliance. This produced the result which claimed that fifty-two per cent supported ANZUS with nuclear ship visits as against forty-four per cent who would rather quit the alliance.

John Henderson, the head of the Prime Minister's Office, was a political scientist. He was certain that combining results from separate questions in the way the committee had was methodologically unsound. It made no allowance for an increase in the margin of error; it forced only part of a sample to a choice. I wrote to the committee and pointed this out. The only result was that the correspondence between me and the committee was published as part of its final report and its findings were launched in a cloud of acrimony.

This was a pity, because its research findings were valuable. The opinion poll it conducted was, setting aside the one point of disagreement, the most comprehensive guide we had ever had to public attitudes to defence and security. One task the committee had set itself was to discover if there was any consensus about defence issues. It seemed that there could be, but it was consensus based on what might be rather than what was. Something like eighty per cent of the population would have been happy to see New Zealand in an operational ANZUS alliance if the United States accepted New Zealand's nuclear-free status.

Manila made it clear to the committee, if it hadn't been clear before, that this appealing option was simply not open to us.

The committee looked for other grounds for agreement. Support for New Zealand's membership of alliances was very high. Over eighty per cent of the people polled believed we should form alliances with other countries. Seventy per cent supported our membership of ANZUS. Indeed, its atavistic allure had hardly diminished. Asked if New Zealand was safer inside or outside ANZUS, seventy per cent replied that we were safer inside, in spite of the fact that very few of the public objectively believed that New Zealand's security was in any immediate sense threatened. When asked to rank the perils facing us, the people polled thought we were most at risk of having our fisheries poached and least at risk of being invaded by enemy force.

Into this fondness for alliances was injected almost as great a dose of nuclear-free sentiment. The committee's response was pragmatic; it reported that public feeling about defence would best be served if the government were to expand New Zealand's relationship with Australia, under the banner of the ANZUS treaty.

I saw a lot of merit in this. Concentrating on regional security issues would, I believed, eventually resolve one of the continuing difficulties of defence policy-making in New Zealand, which was that nobody took our armed forces seriously. To earlier governments, they were the price we paid for American protection. To the public, their function was largely irrelevant. If the armed forces, instead of looking like a set of spare parts, actually looked as if they'd be useful in dealing with troubles in the South Pacific region, then their value to government and public would grow immediately. To the government, they would be a proper tool of foreign policy. In the eyes of the public, their obvious fitness for regional service in the South Pacific would add to the national sense of competence. Concentrating on the relationship with Australia was to me a necessary step on the road from dependence to self-reliance.

The findings of the committee of enquiry were in one sense a vindication. Support for the banning of nuclear vessels was steadily gaining ground. When the committee reported, New

Zealand had been nuclear-free for more than two years. Each day that passed with the country unmolested by the Soviet threat cemented the policy deeper in the public consciousness. At first, support for the policy was very closely linked to patterns of political support. If you voted Labour, you supported the nuclear-free policy because that was part of being Labour. In time, this pattern altered. After the 1987 general election, the popularity of the Labour Government shrivelled. Support for the nuclear-free policy didn't follow it downwards because by then, nuclear-free New Zealand had taken on a life of its own. In 1989, when it was getting hard to find anyone who'd admit to being a Labour voter, over eighty per cent of the population declared themselves to be in favour of the nuclear-free policy. Those numbers spoke volumes. There wasn't any going backwards.

12

Making the Region Safe

NO NUCLEAR-FREE POLICY will last the distance unless people genuinely feel safe in the absence of a nuclear defence. Security in New Zealand's case came down to being able to cope with what was happening in the South Pacific. The military coup in Fiji convinced me that New Zealand's armed forces weren't up to much when it came to reassuring the public of our ability to deal with regional disturbances.

Fiji was a country torn by racial differences. After it gained its independence from the United Kingdom, political power remained for the most part firmly in the hands of the native Fijian population. The majority of the population, by a slight margin, was Indian, the descendants of people brought to Fiji by the British to work in the cane fields. In 1987 the Labour Party, led by a native Fijian, Dr Timoci Bavadra, and relying for its support on poorer Fijians and the Indian community, was elected to office. Like the New Zealand Labour Party, the Fijian Labour Party had a nuclear-free policy, which it immediately implemented. Fiji had an army of a size out of all proportion to its tiny population, and consisting entirely of native Fijians. A few weeks after the Labour Party entered office, army units took its leading members captive at gunpoint within Parliament House while Parliament was in session. They locked them up and installed first a military and then a puppet civilian government.

The coup proved bloodless, but its early days were times of great uncertainty. Speculation raged in New Zealand that our armed forces might intervene to restore the legitimate government, a possibility I never considered. This was in one sense fortunate: our armed forces were quite incapable of defeating the more numerous and well-trained Fijian army in open combat. But I thought it important to convey to the public that we had at least the capacity to rescue our own citizens if the need arose.

I summoned military commanders to my office. I wanted a task force. I wanted aircraft put on standby. I wanted a ship to sail to Fiji to carry home the rescued. The chiefs of the army and air force complied with my requests. I asked the Chief of Naval Staff to ready for sail HMNZS *Monowai*, a survey craft. The admiral was not of a mind to send a ship and urged caution. I told him I was not asking him to commit his craft to combat; as far as I knew Fiji didn't have a navy or an air force to attack it. I asked the Chief of Defence Staff to implement certain evacuation plans. I was advised that he could not, because the Chief of Naval Staff was opposed to the move. I put my instructions in writing. These were conveyed to the Chief of Naval Staff. The admiral refused the instruction. He insisted that my instruction be referred to the Defence Council, as in law he was entitled to do. The Defence Council was a conclave of the government's senior military advisers and the Minister of Defence sat at its head. Its members were summoned and I went to meet them. Here again the admiral balked. He told the council that the *Monowai* would take five days to sail to Fiji. In its former life before it joined the navy the *Monowai* was a banana boat, in which humble capacity it used to make the journey between Auckland and Suva, capital of Fiji, in three days. The admiral was sent off to get another figure and finally came back with an estimate of sixty-seven hours. The *Monowai* was despatched but half an hour out of Auckland its propeller went out of alignment and it had to turn back. One of our frigates made it to Fiji, thanks to the Australians, who refuelled it on the journey.

I knew then that getting our armed forces into useful order would be the work of many years.

In our path were many obstacles. Our military establishment had its own deep-rooted culture. It was bitterly resentful of the challenge presented to its most treasured assumptions by the nuclear-free policy. As time passed, new generations of personnel would learn to take the nuclear-free policy for granted. In the meanwhile, the armed forces were living with the horror of seeing their *raison d'être* swept away from them. Their response was to cling the more firmly to the past.

Another difficulty was that the military establishment was

used to running itself. When the National Party was in government, the position of Minister of Defence was largely ceremonial; every now and then the armed forces would trouble the minister with a request for capital expenditure. That apart, the job was a sinecure. The armed forces were not ready for a minister who wanted to be more than a cipher. Frank O'Flynn, who became Minister of Defence in 1984, thought that some items of defence expenditure, overseas postings and the like, were matters of political and not military judgment. Given the depth of the opposition to the government in the ranks of the defence establishment, I'm sure he was right to be wary of military decision-making. He insisted on being informed, and on exercising his right of intervention. The outcome of his activism was a kind of bureaucratic guerrilla warfare. Papers were lost, reports misplaced and items of expenditure unaccountably overlooked. None of this helped my confidence in our military management.

What I always found hard to swallow was military self-importance. When disaffected officers went to the news media with tales of government foolishness and New Zealand's resulting defencelessness, they invariably presented their complaints as the unbiased assessment of those whose only concern was their country's security. Others justified what in any other walk of life would be seen as self-indulgence with spurious appeals to the morale of the forces.

An example occurred after the death of an admiral who had expressed a wish to be buried at sea. Nothing less than a frigate would do for the purpose. Of our fleet of four frigates, two were possibilities. One was on its way back to Devonport naval base from Singapore. HMNZS *Wellington* was in Australia for calibration of its guns. It was supposed to go from Australia to Apia in Western Samoa, but instead was ordered by its commanders to return to Devonport. Its crew were ashore in Brisbane. With great difficulty, as many of them as could be found were got back on board. The calibration was cut short. To give the admiral's coffin a proper send-off, the crew of the *Wellington* borrowed a kind of launching platform from the Australians. Then the frigate set off at full speed across the Tasman, arriving at the same time as the ship from Singapore. Twelve of its crew missed the sailing and had to be flown over on a commercial flight. Then an air force C130 had to fly across the Tasman to take the borrowed gear back to the Australians. Fully rigged state funerals have cost the taxpayers of New Zealand less.

Fortunately for the public peace of mind, our armed forces were never called upon to deal with any serious crisis. The South Pacific was, essentially, peaceable. This was the assessment the Ministry of Foreign Affairs sent to Cabinet in the immediate wake of the *Buchanan* crisis. 'The government's policy on nuclear ship visits has not changed the security risks to New Zealand. There is currently *no* identifiable threat to New Zealand.' Nobody, however, was going to accuse the government's advisers of letting down their guard. The ministry's analysis read like this: 'What has changed is the way in which ANZUS has operated up till now in terms of military and intelligence co-operation. That change is irreversible, barring a modification of policy on the part either of the United States or New Zealand. Accordingly, we will have to do more, on our own, and bilaterally, with friends and allies, to safeguard our security interests in the South Pacific . . . it cannot be assumed that no new threat will emerge.'

Fortunately the answer was at hand. '. . . While New Zealand's ultimate security rests with the guarantees provided by the ANZUS treaty' (the ministry's thinking on this point was unreconstructed) 'the immediate and practical requirement for us is to ensure, by our own actions and through co-operation with our neighbours, including Australia, that the South Pacific area does not become an area of instability and conflict . . . a new fabric of co-operation needs to be developed with both our South Pacific neighbours and our ANZUS partners — based on a clear definition of New Zealand interests and a policy of self-reliance.'

Cabinet accepted this assessment, and set the government's officials to work out what it actually meant in practice. What it got from the Defence Council was a shopping list. Its report to Cabinet began by noting that unquestioning dependence on the alliance relationship had in the past muffled serious analysis of the defence and strategic issues facing New Zealand. That was the softener. The crunch came next. 'As a result there has been too low a level of national commitment to the provision of armed forces sufficient to ensure independent defence and foreign policies.' The Cabinet then learned that it could remedy this defect by spending more on spare parts and ammunition, raising the numbers in the army, making alternative arrangements for training and exercise overseas, modernising our aircraft, buying a tanker for the navy . . . No itch went unscratched.

I wasn't sure if the defence establishment was trying to frighten us out of the nuclear-free policy or if it was simply looking for enough hardware to set off the gold braid. Its senior officers seemed terrified by the thought that the government would reduce the armed services to a form of civil defence force, limiting their activities to coastal surveillance, fisheries patrols and disaster relief. I didn't think the country was ready for that, but at the same time, it was difficult to take military advice at its face value. To put demands for military resources into their proper perspective, the Cabinet launched the defence review. Our hope was to establish clearly the aims of defence policy in the post-*Buchanan* world, and having done that, match our armed forces to it. The review took nearly two years to complete. It began with an interim review in May 1985, took in the report of the Defence Committee of Enquiry in August 1986, and was finally published in February 1987. Its length was short. It consisted of thirty-eight pages, and two of those pages were occupied by a map of the Pacific hemisphere. But its brevity did not reflect the effort that went into it. After countless meetings of the government's official advisers, and much banging together of heads, the review's broad framework reflected as much agreement as was possible to reach. It was basically a restatement of the position taken by the Ministry of Foreign Affairs early in 1985. We should be more self-reliant and we should look for greater military co-operation with Australia. In the political climate of the time, the review was what the government wanted. It sounded reasonable and practical. But the detail was still left to the never-ending contests of bureaucracy.

More demanding even than bureaucrats were the Australians. Our military links went back many years. When the United States suspended its military co-operation with New Zealand, the Australian Minister of Defence, Kim Beazley, announced that in future his country's armed forces would conduct separate exercises with the forces of New Zealand and the United States. This duplication of effort undoubtedly cost the Australians, and members of their government understandably never failed to remind us of this. Foreign Minister Bill Hayden once spelled it out publicly: if New Zealand wanted an enhanced relationship with Australia it would have to raise its defence expenditure. In private, Beazley gave us just as firmly the same blunt message.

They had us over a barrel. Having put so much of the government's credibility into the defence relationship with Australia, we had to pay the asking price. What the Australians wanted was a major New Zealand investment in military resources. They wanted a New Zealand purchase that would be in effect an enhancement of their own defensive capabilities. Precisely what they wanted was ships.

Our defence review identified a need for a new long-range patrol and surveillance vessel to replace the navy's elderly frigates. It referred to discussions that were already in progress with the Australians about the joint development of a suitable craft. Not long after the review was published, the Cabinet adopted in principle a proposal from its military advisers that the frigates be replaced with what were then called 'seaworthy ocean combat craft'. There were more exploratory talks with the Australians.

Once we got the word that the frigates were the price of Australian goodwill, it became a matter of extracting the best possible deal. We mused publicly about alternative purchases. We acted coy about making up our minds. We haggled over details. But in the end we signed on the dotted line and bought two Australian ships.

To say the purchase wasn't popular would be an understatement. Nobody took much exception to the craft described in the defence review, but when idea got closer to reality and the ships took on more and more of the characteristics of the frigates the Australians wanted to build, exception was taken by the bucketful. In the recessionary economic climate of 1988 and 1989 the expensive vessels seemed like pointless extravagance. They looked as idle and redundant as the vessels they were replacing.

To try to quiet the clamour, the government put a lid on its spending on defence. We set management consultants on to the defence establishment in search of greater efficiency. Little came of our efforts and the frigates remained a major political liability.

If we didn't go ahead with the purchase, I was sure that the Australian Government would greatly reduce military co-operation, leaving us unable to meet the targets set publicly in the defence review. At the same time we were discussing the frigate purchase we were negotiating an extension of the free trade agreement between Australia and New Zealand. I didn't want that derailed by a fit of Australian pique about the purchase. The frills that appeared on our 'seaworthy ocean combat craft' were a price I was willing to pay for that.

Security in the South Pacific meant more than an occasional show of usefulness by New Zealand's armed forces — it meant the continuing absence of unsettling outside influences. It meant keeping the dangerous and destabilising confrontation between the superpowers out of the region. With that in mind I went to the 1984 meeting of the South Pacific Forum to resurrect the idea of a South Pacific nuclear-free zone.

The member states of the South Pacific Forum were the tiny island countries of the South Pacific, along with New Zealand, Australia, Papua New Guinea and Fiji. Each year their heads of government met, far less formally than at any other heads of government meeting I ever attended, to discuss matters of moment to the region and, in the time-honoured Pacific way, allow a consensus to emerge slowly from the talking. The 1984 forum was held in Tuvalu.

Tuvalu consisted of a number of palm-fringed coral atolls, chief among them Funafuti. Its airstrip covered about a third of its land area. The only car on the island belonged to the Governor-General. I was provided by the islanders with a bicycle which carried the wording 'Prison One'. On this little atoll descended the military might of Australia and New Zealand, carrying with it the supplies and equipment that made possible the holding of an international conference in a place otherwise totally unprepared for it. I was living in a hut on the edge of the airstrip. There was a kettle, a sink and some running water. I wanted tea; but there wasn't any. I asked my ever-resourceful private secretary if he could get some for me. He had armed himself with a two-way radio. Out in the lagoon floated HMNZS *Monowai*. Now I heard my secretary say, 'Camp David calling *Monowai*, Camp David calling *Monowai*.' Before I could take exception to this extraordinary description, he added, 'The Prime Minister wants some tea.' Minutes passed, then the rumble of a powerful engine was heard. A three-ton truck that had been flown up from New Zealand for the conference pulled up outside. In jarring fashion a soldier saluted and presented me with a teabag. The good folk of Funafuti were agog, but then so was I.

Every international conference of any pretension to seriousness must have what is called a retreat. Here the leaders of delegations seclude themselves for private talks. On Funafuti, the heads of government took themselves off to an even smaller island in the middle of the lagoon, where models of ships' cabins and dining rooms had been built. Many of the menfolk of Tuvalu earned their living as stewards on merchant ships, and on the little island they learnt how to make up berths and set tables in the approved style of the major shipping companies. I was sure that in the mock-up of a dining room I would see South Pacific politics at its best.

The conversation was desultory. The President of Nauru fell asleep, cigarette in his mouth. When it burnt to his lips he woke and said to Australia's representative, 'Prime Minister Hawke, can I ask you about that man in Melbourne who is ruining my shipping company?'

Hawke said, 'It's all right, mate, it's all been fixed, it's all right.'

Reassured, the President lit another cigarette and went back to sleep. When it in turn burned down and woke him up, he asked Hawke exactly the same question. Hawke gave him exactly the same reply.

The fourth time it happened, Hawke stood up in his place and screamed, 'I'm telling you the cunt's dead!' Conversation became even more stilted after that.

In formal assembly, the forum agreed in principle to the establishment of a South Pacific nuclear-free zone. We set up a working party of officials whose task it was to draw up a draft treaty which the heads of government would examine at their meeting the following year at Rarotonga in the Cook Islands. There was little or no discussion of New Zealand's nuclear-free policy, but then there was actually little discussion of any country's policy. This was just as well, given the differences between us. New Zealand and Vanuatu were nuclear-free. Australia and Fiji accepted visits from American nuclear ships. Australia also played host to elements of the American system which gave early warning against nuclear attack. There was never any possibility of reconciling those differences and it was understood from the start that the limits of the nuclear-free zone would be drawn at what we *could* agree on.

The treaty signed at Rarotonga on Hiroshima Day 1985 obliged each of its signatories not to manufacture nuclear weapons. Nor were nuclear weapons to be acquired in any other way. Each signatory undertook not to station nuclear weapons on its territory (which was not the same as allowing nuclear weapons to pass through on visiting ships and aircraft). Testing was outlawed, and so was the dumping of radioactive wastes.

Attached to the treaty were protocols seeking the endorsement of countries outside the South Pacific region. The first protocol invited the United Kingdom, France and the United States to accept the obligations of the treaty on behalf of their dependent territories in the South Pacific. If France adopted this protocol, it would be obliged to put a stop to its nuclear testing programme at Moruroa. (France didn't, of course.) A second protocol invited the nuclear powers to refrain from testing nuclear weapons in or on the high seas of the South Pacific. (The United Kingdom, the United States and the Soviet Union were already bound by international treaty not to test there. China and France were not.) A third protocol invited the nuclear powers to give to the signatories of the treaty an unconditional promise not to use or threaten the use of nuclear weapons against any one of them.

The treaty's limitations were obvious. By international law, all vessels, including nuclear-armed and -powered vessels, may pass freely through international waters. Equally, aircraft armed with nuclear weapons must be allowed to pass freely through international airspace. The Treaty of Rarotonga had to recognise this. It would have been possible to add a protocol to the treaty which invited the nuclear powers to keep their portable nuclear arsenals out of the South Pacific, but this possibility was never seriously considered. It would have killed whatever chance there was that the nuclear powers would put their names to the other three protocols. The signatcries to the treaty made no commitment to exclude nuclear-capable ships and aircraft from their territory. Had such an undertaking been required, Australia wouldn't have signed and we wouldn't have had a treaty.

Australia in fact was an anomaly. Unlike New Zealand, it still enjoyed the security guarantee given by the United States under the ANZUS alliance. If this guarantee meant anything at all, it meant that the United States was willing to use or threaten to use nuclear weapons to deter an attack against Australia. The American bases at North-West Cape, Pine Gap and Nurrangar the Australians acknowledged to be likely Soviet nuclear targets, their credentials as bearers of the burden of deterrence. It was plain that the Australians did not take the South Pacific nuclear-free zone too seriously. Indeed, Bob Hawke's anxiety at Tuvalu to claim paternity of it suggested that its appeal lay firmly in the realms of political expediency. It sounded good, but it didn't stop the Australian Government doing what it wanted, and it didn't put any immediate limits on the activities of its American ally. A deputy assistant secretary of state told our ambassador in Washington in 1987 that there wasn't a non-nuclear South Pacific and the nuclear-free zone didn't create one. I don't know if this was a sneer or an expression of relief.

Whatever its private feelings, the United States chose to be publicly disapproving of the nuclear-free zone. It refused to ratify the protocols, saying that in view of what it called its global security interests and responsibilities, it was unable to endorse them. The Soviet Union signed the protocols. So we ended up shunned by the power with which most of the South Pacific identified and embraced by the one with which we didn't.

Beside the limitations of the South Pacific nuclear-free zone must be set some advantages. It had the real merit of prohibiting the stationing of nuclear weapons in the region. We weren't going to end up with an American naval base in Fiji or nuclear submarines home-porting at Cockburn Sound in Western Australia. This act of self-control became meaningful in light of the continuing American search for facilities to replace the Philippines when the future of its bases there was put in doubt. In imposing on themselves this genuine restraint, the countries of the South Pacific, for all their differences, showed the political will that is ultimately our only guarantee against the deployment and use of nuclear weapons. It was a start, and you surely have to start somewhere.

I think it was a Secretary-General of NATO who complained about the concept of nuclear-free zones by saying that if the people of the United States were left by themselves to carry the burden of deterrence, they'd get tired of it. I couldn't have put it better myself. He was exactly right: no country should carry the burden of defending others with weapons that can destroy the planet; the others should manage to look after themselves. We might in the South Pacific have armed forces that don't meet NATO standards, but that's not what it takes in this part of the world. The South Pacific can look after itself, and I'd like the nuclear powers to lay down the burden of defending us.

13

A Popular Platform

CONTROVERSY RAGED OVER THE USS *Buchanan*. The State Department thought of new ways to insult our diplomats. George Shultz took away the nuclear umbrella from above our heads. All the while the United States Navy stood guard at its last outpost in New Zealand.

The navy had a base at Harewood near Christchurch. It was nowhere near the water, but stood on the boundary of the airport. Harewood had been for many years the support base for the Antarctic operations of this respected American scientific body, the National Science Foundation. The foundation relied on the navy for transport, so to Harewood came military transport aircraft on their way to and from the ice. To Harewood also came military transport aircraft on their way to and from Australia. This inoffensive operation I was happy to see continue. Like the Royal Yacht *Britannia*, the Starlifter aircraft which made their way to Christchurch were shrouded in the ambiguity of 'neither confirm nor deny'. But I was in my own mind certain that the aircraft landing at Harewood were innocent of nuclear weapons. The Americans did not tell me this; like the owners of the *Britannia*, the owners of the Starlifters wished their property to lose none of its mystery.

The fact of it was that nuclear weapons were not tossed lightly into the hold of transport aircraft. Careful safeguards against accident must, by American military protocol, be in place. The same protocol demanded that the weapons and their component parts be guarded against the risk of theft. There was no doubting the strength of American concern about the spread of nuclear weapons to countries that were not already nuclear powers. That concern was reflected in precautions routinely taken against their violent seizure, and of such precautions there was no sign at Harewood. The only thing the Americans were trying to hide was marijuana, which was sometimes smuggled on board their aircraft. This low-level threat we could probably live with.

Members of the peace movement regarded the Harewood base with deep suspicion. I don't recall them expressing reservations about the Antarctic flights; Antarctica was the subject of a treaty, which the United States had signed, excluding nuclear weapons. It was the flights to and from Australia that made them anxious. Perhaps the *Britannia* would have made them anxious. I didn't share their worry. It never seemed to me that our nuclear-free policy required that we demand from the Americans an accounting of the contents of the hold of every Starlifter — we made our own judgment. We had made a judgment long before about the American military aircraft that came to New Zealand for the Triad exercise: they weren't carrying nuclear weapons. The USS *Buchanan* defeated our resources and, since we didn't know what it was carrying, it had to stay away. In cases where we were capable of drawing our own conclusions, it didn't matter that the Americans remained silent.

Like foreign warships visiting our ports, foreign military aircraft entering New Zealand airspace first had to obtain the permission of the government, and I granted permission to the Harewood aircraft. Then the nuclear-free bill made its appearance in Parliament. One clause provided that the Prime Minister would not grant permission for the entry of foreign military aircraft without being satisfied that the aircraft were not carrying nuclear weapons. On the face of it, the clause should have caused us as many difficulties as that dealing on the same basis with warships. It was no less and no more a challenge to the doctrine of 'neither confirm nor deny' than was the ship clause. In the event the trouble didn't eventuate. Fighting aircraft of the United States were no longer seen in New Zealand, but the Starlifters kept right on coming to Harewood.

The apparent reason for the difference was a refinement of the wording of the aircraft clause. Unlike the ship clause, it allowed the Prime Minister to give approval, not merely to individual aircraft, but to categories or classes of aircraft. The approval could cover repeated visits by aircraft of a given class or category. Even before the bill became law I approved on this blanket basis the use of Christchurch airport by the transport aircraft, an approval the Americans chose to regard as unconditional. Because they didn't have to get clearance for each individual flight, they could assert to themselves and anyone else who cared to listen that the New Zealand Government was indifferent to the cargo carried in the aircraft. Starlifters with all the menace of *Britannia* continued to land at Christchurch.

The Americans could not seriously have expected any challenge to their aircraft. They didn't have any point to prove at Harewood, and, having convinced themselves that the wording of the legislation was enough to save face, they didn't have any cause to stay away.

This was my understanding of what was happening. The State Department wouldn't set itself to boasting that New Zealand had succumbed to the demands of America's nuclear strategy, while I wouldn't go round declaiming that the United States was complying with New Zealand's nuclear-free policy. The understanding never amounted to any kind of agreement, either formal or informal: the State Department used to take strong exception to the least suggestion that New Zealand and the United States had reached any kind of accommodation over Harewood. Just as I didn't want to be seen as surrendering to them, they didn't want to be seen as complying with the nuclear-free policy. With this agreement not to agree I was pleased enough.

I am not saying that Harewood's future was never in doubt. Even before the Labour Party became the government, H. Monroe Browne used to mutter that the Harewood operation would be moved to Hobart in Tasmania if the nuclear-free policy made life here too difficult. This would represent a not-insignificant loss to the job market and the retail trade of Christchurch. After we turned away USS *Buchanan*, the United States Navy pressed for the removal of the base as a form of punishment. Navy Secretary John Lehman said on his retirement in 1987 that 'Our position in the Navy is to make it very clear that they must understand that they cannot expect to enjoy special privileges and relationships, if they choose to buy off their left-wing extremists by kicking the United States Navy in the teeth.' Lehman added that planning for the removal of the base had already started; the navy was concerned that the passage of the government's nuclear-free legislation would, as he put it, challenge operations at the base.

Staff at our embassy in Washington scurried around the bureaucratic traps to find out if there was anything serious in Lehman's latest outburst, but little new emerged. As always, the navy wanted the base moved. As always, it was unable to win support for the move from the rest of official Washington. The State Department did not wish to see the dispute over ANZUS spread to scientific co-operation between the United States and New Zealand in the Antarctic. Having lost the bureaucratic battle, the navy was now trying to frighten official Washington with speculation as to what might happen when the nuclear-free bill finally became law and what befell the *Buchanan* happened instead to one of its aircraft.

A State Department official advised our diplomats that it was not in the American interest to leave Harewood as long as 'there is no issue posed by the New Zealand side', which we translated as meaning that Harewood would stay until such time as the government actually refused entry to an American military transport aircraft. This convenient arrangement was threatened by the suspicion of the peace groups. They remained convinced of the sinister content of the Starlifters bound for Australia. Arguing that the government was turning a blind eye to American activity at Harewood, they attempted to rouse public concern to the heights it had reached over the *Buchanan* visit.

In this attempt they were aided by the Leader of the Opposition. After the 1984 election, Robert Muldoon lost what remained of his standing among his parliamentary colleagues. In a paroxysm of repentance, the National Party caucus elected a liberal product of the urban professional classes to lead it. But it was not at heart a liberal party and its new leader proved himself out of touch with the feelings of his party's traditional following. The party soon went back to its roots, choosing as its next leader a man with the bearing of a simple country lad. This was Jim Bolger, whose only distinction was his curious habit of referring to himself in the third person. This scion of rural conservatism could not have been moved by the prospect that the Harewood

Starlifters might carry nuclear weapons. Given his party's steady support for American policy, he should have been reassured by the thought that deterrence was on the job in Christchurch. His aim of course was not to safeguard the purity of the government's nuclear-free policy, but to challenge the government with inconsistency.

In Parliament Bolger accused the government of taking one approach to ships and another to aircraft. Waving an official report of a meeting he had in Washington the year before with a senior official of the Defense Department, he hinted darkly that the nuclear-free policy had been sold down the river to pay for the base at Harewood. Tackling these accusations wasn't always easy with an arm tied behind my back. I didn't think I could do more than say that the Harewood aircraft came to New Zealand in accordance with the government's policy. If I said it was the Americans who blinked, the United States would have gone off in dudgeon, taking Harewood with it. If I said that when the nuclear free bill became law my decisions about the aircraft would be subject to judicial review, there would have been the same result. I wasn't very happy about having to hedge, but it seemed silly to lose an inoffensive operation like Harewood to score rhetorical points.

If the peace movement and the parliamentary Opposition had succeeded in frightening the public about Harewood, I would have had to take a different tack. Had doubt about Harewood led to a loss of public confidence in the nuclear-free policy, then either I had to embarrass the United States into closing Harewood, or close it down myself. Fortunately, it didn't happen. Most folks used their common sense, and a Harewood scare never took off. In June 1987 the nuclear-free bill finally became law and Harewood stayed.

In one respect the Leader of the Opposition was as handicapped in his efforts as I was. He could attack the government for its supposed inconsistency about Harewood, but he couldn't base his appeal on the undoubted public aversion to nuclear weapons. He might have caused us far more trouble than he actually did if he'd been coming at us armed with contempt for deterrence and all its works. But locked as he was into the

American perspective, he couldn't wrest from the government the mantle of defender of nuclear-free New Zealand. It was the government and not the Opposition which went into the general election of 1987 on the side of the angels.

On the side of the National Party was the Western alliance. American Ambassador Paul Cleveland took to the rubber-chicken circuit to plead the virtues of the bomb, preaching the value of deterrence wherever he could find an audience. He had about as much success as the effort to christen the MX missile the Peacemaker. He never struck the right note. To most New Zealand audiences deterrence was essentially foreign, it didn't belong here.

Nor did many people care for overt attempts by foreign powers to change the nuclear-free policy by changing the government. Easily the most inept of those interventions was made in April 1987 by the British Foreign Secretary, Sir Geoffrey Howe. This amiable minister must have been seen as the last throw of the dice by the increasingly desperate British High Commissioner. In Canberra on his way here, Howe suggested that New Zealand was enjoying a 'free lunch', benefiting from the efforts of the Western alliance without paying its share of their cost. In New Zealand, we heard that there was no hiding place from ever-expanding Soviet interests, and that we were letting the side down by not allowing ourselves to be defended by nuclear weapons. But most of the publicity that surrounded Howe's visit came from his raising of an old bogey, our continuing struggle to sell our butter in Britain in face of the protests of the subsidised butter-producers of the European Community.

This is how he put it: 'With your current defence policy, it is a fact of life that your cause is less likely to prevail in a European Community, eleven of whose twelve members also belong to NATO.' Translated into headlines, this emerged as 'Nuke ban threatens trade says Howe'. Either way, it didn't run. By 1987, people had worked out that there wasn't anything very principled about European farm politics; whatever was happening to our butter didn't have a lot to do with the solidarity of NATO. Sir Geoffrey's intervention must have helped the government far more than it hindered us.

Howe was a good-natured fellow. I met him, after the election, in Vancouver, at the Commonwealth heads of government conference. He and I didn't look much alike, but neither of us was slender and we both wore glasses. Now he said to me, 'It's been a good day for you, Prime Minister.' I didn't understand what he meant. 'Four African leaders,' he continued, 'have congratulated me on my government's anti-nuclear policy.' He said he hoped I didn't think that his visit to New Zealand had been unnecessarily unhelpful to me in my election campaign. 'Not at all,' I said, 'you come back in 1990.' Unlike the State Department, Howe could laugh about it. In fact the Labour Party's polling showed the value of the nuclear-free policy in attracting and retaining voter support. It won Labour support among members of the public who found the government's economic policies hard to swallow. And it put the National Party in a lather of indecision. The National Party's strategy was always to focus on the value to New Zealand of the ANZUS alliance. Their stated aim was to return New Zealand to what it called 'operational membership' of the alliance, by which it meant an alliance undisturbed by speculation about the weaponry carried on American ships. In the early days of our dispute with the United States, the parliamentary Opposition was adamant that it would restore us to the alliance by accepting without question the visiting ships and aircraft of the country's allies. In May 1986 Jim Bolger acknowledged that National's acceptance of 'neither confirm nor deny' meant that nuclear-armed ships could enter New Zealand ports. He was reported as saying that he did not believe that National's policy would hamper its wish to project itself as a peace party.

By the time of the election it must have been apparent to National Party strategists that its support for ANZUS was too easily translating in the public mind into support for nuclear weapons. I certainly never missed an opportunity to remind people of the connection. Not long before the campaign started, Bolger tried a new option. Still refusing to challenge the policy of 'neither confirm nor deny', he said that a National Government would not want its allies to send nuclear-armed ships to New Zealand. His party would, he said, in contrast to the Labour

Government, trust its allies to respect New Zealand's anti-nuclear wishes. This effort to reconcile a nuclear-free policy with active participation in ANZUS I later described as the National Party's quest for the Holy Grail of New Zealand diplomacy, which is probably a libel on the knights of the Round Table.

The new refinement did Bolger no good. Former Chief of Defence Staff Ewan Jamieson was reported as saying that he would support a policy stating that New Zealand did not want nuclear ship visits, while trusting allied governments to respect that wish. It was what countries like Norway did. 'But they do not implement that policy in a way which prevents their allies from coming into their ports. That is where we have been different from anyone else.' In terms of New Zealand politics, the endorsement of the former Chief of Defence Staff was as good as a seal of approval from the Pentagon. It amounted to saying that we could trust our allies because we knew they weren't to be trusted. But in 1987 the distorted logic of deterrence cut no ice in New Zealand. For all the National Party's attempts to wrap deterrence up in the flag, the alliance of nations and the sacrifices of the past, it couldn't get around the fact that its policy invited New Zealand to applaud while the nuclear powers gambled with the future of all the rest of us.

I gave only one speech about ANZUS and the nuclear-free policy during the election campaign. It was in its way an account of the journey we'd been on since 1984. I started by remembering that the Labour Party had campaigned in 1984 on an undertaking to renegotiate ANZUS. 'There was certainly no intention of leaving the alliance or becoming a sleeping partner in it, and when I was campaigning in that election I was assertive of the value to New Zealand of the alliance.' At that time, I recalled, I believed that New Zealand could exclude nuclear weapons and remain in active alliance with a nuclear power. I didn't see the alliance as predominantly nuclear, but events proved me wrong. The alliance was a vehicle of nuclear strategy. 'The ANZUS relationship between the United States and New Zealand is now inoperative exactly because the nuclear element in the alliance has become predominant.' After describing the many efforts which had been made to reconcile the irreconcilable, I concluded by saying that ANZUS had been unequivocably revealed in the last three years to be a defence arrangement underpinned by a global strategy of nuclear deterrence. 'As long as it retains that character, it is no use to New Zealand and New Zealand had better make arrangements which are relevant to our own circumstances.'

The National Party paid me a marvellous compliment. Seeing my speech as an epitaph for ANZUS, they turned it into a television advertisement. I've never really thanked them for it.

My speech also evoked a more traditional response, a complaint from the Americans. The United States did not care to be characterised as I had characterised it, the inevitable bearer of nuclear weapons to the ports of its allies. Too bad, I thought.

What I never lightly dismissed was the value of the nuclear-free policy in keeping up the spirits of the Labour Party. In the 1987 general election, it had a unifying effect which had no parallel in any other policy we ever adopted. Its appeal to Labour people was unalloyed, the kind of appeal that made it a pleasure to take the political battle to the party's opponents. The same could not be said of other policies adopted by the Labour Government. Barely a few months after the 1984 election, differences over economic policy emerged between the government and the party outside Parliament. In clearing away the economic controls of the Muldoon era, the government left itself open to the accusation of the wider party that it had abandoned the egalitarianism which was the essential characteristic of the labour movement. Our differences gave rise to the slur that the nuclear-free policy was a sop thrown by the government to the party at large. It was, in other words, alleged to be the price paid by the government for the party's acquiescence in its economic management.

It must be acknowledged that many outsiders were genuinely puzzled by the fact that a government which was in most ways a model of capitalist orthodoxy should adopt a foreign policy which some in the United States and Europe saw as a comfort to advancing communism. Some kind of trade-off between government and party was all they could think of to account for it. The explanation doesn't fit the history. The nuclear-free policy was a fact of life in the Labour Party long before government and party took to fighting over economic policy; left and right alike supported it. Even after 1987, when disenchantment deepened and economic policy came to divide the parliamentary caucus as much as it had once divided government and party outside Parliament, I couldn't place people's loyalties according to their view of the nuclear-free policy.

We were lucky in 1987. The fatal divisions between left and right in the government had hardly reached the surface. The sharemarket roared and people felt confident. On election night the Labour Party increased its majority in Parliament. The American ambassador did not send his congratulations.

14

Fool's Gold

AFTER THE GENERAL ELECTION, I gave up the foreign affairs portfolio. I knew that the foreign policy crises of the government's first three years in office were unlikely to be repeated. I thought it would be useful to have a period of consolidation, when we turned down the volume and allowed the dust to settle. The new Minister of Foreign Affairs was Russell Marshall, whom I chose from among the members of Cabinet elected by the Labour Party's parliamentary caucus. In our first term of office he was Minister of Education. He was a minister who was not ordinarily inclined to set up in opposition to the advice he received from his official advisers, and in that sense his appointment offered an opportunity to our diplomatic establishment to reassert itself. At the same time, as much as anybody in the Cabinet, Marshall had a deep-seated dislike of nuclear weapons, and an instinctive distrust of those who tried to justify their deployment. I saw no risk in his appointment to our nuclear-free policy.

Some months after Marshall's appointment, I opened my morning newspaper and read a disconcerting headline. 'Surprise shift in nuclear policy', it announced. The story ran under the by-line of a Wellington journalist. To this reporter the capital's diplomatic community delighted in rehearsing what they saw as their triumphs and the government's embarrassments. Now I read that the government had made a fundamental alteration in its nuclear-free policy. According to the newspaper, we acknowledged that nuclear deterrence had an important role to play in maintaining world peace.

It quoted from a speech by Marshall to the United Nations Conference on Disarmament in Geneva. The minister's reported words seemed to justify the headline. He had gone beyond mere recognition of the fact that deterrence was important to the global strategies of the great powers. Instead, he said that deterrence was keeping the peace at the global level. This could have been written by the State Department. To prove the point, the article reported 'Western diplomats' as saying that Marshall's speech marked a return to normality after what it called the hype and hysteria of the past few years.

A search among the cable traffic between the Ministry of Foreign Affairs and its diplomats abroad showed what had happened. In July 1987, Graham Fortune, New Zealand's ambassador to the Conference on Disarmament, made a speech critical of the assumptions behind nuclear deterrence and called deterrence a paradox. He pointed out that we might be certain, when it was too late, that deterrence had failed, but we could never have absolute proof of its success. 'The idea of "us" deterring "them" has, in the nuclear age, a new and absurd meaning,' the ambassador said. He also put New Zealand's position. 'New Zealand does not accept that Western security must be indivisibly reliant on nuclear weapons. We believe that alternatives to nuclear deterrence do exist.'

In Geneva there was a cottage industry in disarmament and arms control. Many countries were represented at the Conference on Disarmament. When not making speeches themselves, the ambassadors and observers read the speeches of the others. If New Zealand's experience at Geneva was any guide, the ambassadors of the nuclear powers had little time for independent initiatives. Submission to the party line was what was wanted. Displeasure was duly visited on those who did not join the chorus of approval for the actions of the superpowers.

The United States' representative at the Conference on Disarmament was Ambassador Max Friedersdorf. He called on New Zealand's ambassador to express the administration's concern about Fortune's speech. The United States, he said, had been particularly disturbed by passages that appeared to question the usefulness of nuclear deterrence and the indivisibility of Western security. Friedersdorf then read the unfortunate Fortune a lengthy statement, written in Washington.

He began by reminding Fortune about the threat posed to democracy by the Soviet Union: 'The threat of aggression is very real, based on a hostile ideology, significant military capabilities, and concrete actions.' He did not go into detail here. Instead, he reminded Fortune that the security of the West depended on effective military strength, including of course, nuclear strength. Then followed a hymn to freedom, democracy, justice and diversity, and the need to defend these Western values through nuclear deterrence.

Perhaps mindful that their rebuke was to be delivered by the American representative at a conference on disarmament, the authors of the message then set out on a different tack. President Reagan himself had stated that a nuclear war could not be won and must never be fought. 'Over time,' the ambassador intoned, 'the international community must find other means to ensure international security.' But not right now. NATO foreign ministers, the ambassador reminded Fortune, had recently reaffirmed their deterrence policy in the face of the persistent Soviet build-up.

New Zealand, it seemed, was being selfish: 'The statement that the South Pacific can and should remain free of nuclear weapons misses the basic point. The independent nations of the South Pacific are the beneficiaries of the West's nuclear deterrent, which plays an essential role in preserving peace and freedom.' It was time New Zealand understood that it was deterrence which kept our region unmolested by Soviet Russia. Everyone else had got the point: 'Others in the region, especially Australia, have emerged from strategic analyses with quite different conclusions.' The Australian defence white paper accepted the need for collective Western defence; it argued that Australia could only achieve greater self-reliance with American support (this was actually what it did say). The Australian assessment declared Australia willing to take the risks involved in contributing to nuclear deterrence.

Selfishness wasn't our only failing. The ambassador continued, 'In your address you stated that New Zealand does not intend to harm the interests of other Western countries. In fact, by publicly eroding the unity which lies at the heart of successful Western collective security with doubtful arguments about regional divisibility, New Zealand does in effect judge what is best for others and harms Western interests.'

New Zealand did not understand that it was only arms build-ups which led to arms reductions, the ambassador said (he really did). Would the Soviets think about removing their SS-20 missiles if NATO hadn't outplayed them with its deployments of Pershings? The disarmament ambassador ended on a note of condescension, tinged with warning. New Zealand was entitled, he said, to define its own interests at the Conference on Disarmament. He could even understand how 'Prime Minister Lange' might be moved to echo some of those views in the heat of an election campaign. But, he concluded, 'A concerted effort over the long-term to push in Western councils formulations outlined in your address would be seen as distinctively unhelpful in efforts to maintain effective collective security and achieve meaningful progress in arms control.'

Unburdened of this message, the American ambassador rested while New Zealand's representative defended his country's position. Fortune later reported that he believed Ambassador Friedersdorf to be no ideologue; he felt that he had delivered the State Department's message less in anger than in sorrow.

The next day Fortune cabled to Wellington that the Australians had called on him. In milder terms than used by the United States, the Australian representative took issue with Fortune's analysis of deterrence. It was not an approach Australia could endorse, not for the time being at any rate. Of more concern to Canberra was what the Australians saw as the attempt, implicit in the speech, to carry New Zealand's policies to the rest of the South Pacific.

In Washington, Ambassador Bill Rowling was summoned to the State Department to meet a deputy assistant secretary of state for East Asian and Pacific affairs. Rowling, too, was to be told the administration's views on Graham Fortune's speech to the Conference on Disarmament. This came as a surprise to the ambassador, who had often said much the same as Fortune had without attracting particular notice. Now he heard from the State Department's representative that some parts of the speech gave the United States no problems. But when it turned to the New Zealand Government's views on nuclear deterrence and nuclear weapons, it became very troublesome. New Zealand should know better than to talk about the paradox of deterrence. Nuclear weapons weren't only preventing nuclear war, they were preventing conventional war. 'You're just as dead from a bullet as from a nuclear weapon,' the deputy assistant secretary argued, overlooking for the moment the point that a bullet never killed a planet.

Where we'd really gone wrong was in thinking that security in the South Pacific could be separated out from security in Europe. It was that kind of thinking that led to the First World War, and the Second World War (he never explained what he meant by this extraordinary allegation). The whole Western security structure since the Second World War had been created in acknowledgment of the fact that countries could not and would not be allowed to unilaterally pursue their own security arrangements in separate parts of the world.

The deputy assistant secretary's worries didn't end with the possibility that other countries might follow our unspeaking example. He wanted us to stop *talking* about the reasons behind our nuclear-free policy. I suppose that the emperor who had no clothes wanted the little boy to stop chattering.

All of this Bill Rowling heard and reported to Wellington. Finally, to make sure nobody missed the point, Ambassador Paul Cleveland put the same word around the diplomatic traps in Wellington. Then came a variation. When contact was made between New Zealand diplomats and American officials, the Americans held out the carrot of increased diplomatic contact. The stick was already in plain sight. Talking out of turn about deterrence would decrease the chance of friendly dealings.

At a dinner given later by our ambassador in Washington, an assistant secretary of state confirmed that the United States had adopted a new position towards New Zealand since the election. He did not, of course, acknowledge that the result had put an end to any immediate hope that we would surrender the nuclear-free policy. Instead, he resigned himself to the fact that there wasn't going to be a solution to the problem in the foreseeable future. He talked about how much we had in common; above all else, our commitment to democracy. He talked about the disarmament negotiations going on between the superpowers, and how the United States more than ever needed the help and support of its friends and allies. He asked New Zealand to stand alongside the United States. He asked, at the least, for us to do nothing that might weaken the solidarity of the West.

Our representatives in Washington weren't slow to take this on board. The word went back to Wellington. 'There is no doubt that we are being asked for our support of the United States over the period ahead, in a way that will test the government's resolve to improve bilateral relations and our many assurances that we are not drifting off into neutrality or non-alignment.' In other words, we'd have to be seen to take the American side when the two superpowers got down to arms reduction talks. Washington spelt it out for the folks at home. It would not be enough to treat the superpowers evenhandedly, or leave the impression that New Zealand was less than a fully committed member of the Western group. Boots and all was what was wanted. The Ministry of Foreign Affairs in Wellington heard, and understood.

The result was that Russell Marshall went to Geneva and bowed down before the altar of nuclear deterrence. Marshall, it seemed to me, was as taken aback as I was by the reporting of his speech. He had, after all, said what his diplomatic advisers had suggested he say. He had taken their advice that what he said was consistent with our policy, as well as being timely and appropriate. Now we both suffered the embarrassment of having to explain his speech away and record again in public our opposition to the principle of deterrence.

I wrote to Marshall some time later in a formal response to a policy paper he sent me. I said that in the end it didn't help us to make allowances for the thinking behind deterrence, or for dubious doctrines like NATO's refusal to adopt a policy of no first use of nuclear weapons. 'I have found from experience,' I said, 'that any attempt to acknowledge the circumstances which have led to the adoption of these doctrines is inevitably presented here as some sort of capitulation. Nor in fact does it do us any good abroad. An understanding tone makes little difference when our whole practical policy is a repudiation of nuclear deterrence.'

By now New Zealand had a new ambassador in Washington. His name was Tim Francis, once victim of an alsatian bite in his country's service at The Hague. He was a career diplomat who had been at the point of retirement when I asked him to serve in Washington; I enjoyed working with him and had seen him engage in some forthright defences of the government's policy. I trusted him — his was not the inclination to see nuclear vessels anchored once again in New Zealand harbours. He was for all that an indefatigable conciliator. In response to the signals sent out by the American administration, he set out to reconstruct his country's relationship with the United States.

Francis went to present his credentials to the acting Secretary of State, ordinarily a mere formality. To his astonishment, he found himself warmly welcomed. The acting secretary, while noting the sharp differences between New Zealand and the United States on security issues, invited the ambassador to call on the State Department for help on any other matters he might like to raise, trade or tourism and the like. Nor indeed, when pressed, did the acting secretary rule out future co-operation in areas ordinarily deemed out of bounds by the United States, like intelligence. When Francis suggested that we might exchange information about Russian activities in the South Pacific, the acting secretary told him that this possibility could be explored. He then read the ambassador the usual lecture about not exporting our nuclear-free policy. To Francis, that was an invitation to tell him how devoted to the West we truly were.

Lower down the State Department, the signals, to start with, were mixed. An assistant secretary of state told Francis that there would be no letting up on the refusal to exchange intelligence information with New Zealand. If that happened, countries with the same nuclear-free leanings as New Zealand might think they too could get away with it. (But we weren't to worry. He wouldn't let New Zealand come to harm by withholding information directly relevant to our security.) Francis hastened to assure him that the New Zealand Government had expressed strong support for President Reagan's arms control initiatives. We acknowledged the part played by nuclear deterrence.

'If that's the case,' the assistant secretary asked, 'why can't you take the hard decisions?' The NATO countries took Cruise missiles. Why couldn't New Zealand take ships?

He wasn't prepared to listen to any argument that the South Pacific was different. Still less would he listen to Francis's protestations that the nuclear-free policy wasn't for export. 'It's the policy that's the problem as far as I'm concerned,' the assistant secretary said.

While our new ambassador diligently scoured Washington for signs of warmth, the government he represented was still put among the untouchables by official Washington.

In March 1988 the leader of the National Party visited the United States. He was granted a meeting with the Secretary of Defense. No administration official of comparable standing had met any member of the New Zealand Government since I parted company with George Shultz in Manila. With some difficulty, members of my staff persuaded the Ministry of Foreign Affairs in Wellington to point out to the American ambassador that his government's action would be seen as partisan. The Americans took no notice. I thought their conduct wasn't so much partisan as stupid.

These slights directed at the government only made the ministry more determined than ever to repair the damage. Francis pressed on through the ranks of the administration, tirelessly seeking signs of accommodation. He reached the conclusion that the security disagreement, if it couldn't be settled, could at least be set to one side. There would soon be a new administration in power in the United States, and he saw this as the chance for what he called 'a resumption of dialogue at a political level'. There would be changes in personnel in Washington: veterans of the fight over USS *Buchanan* would quit, taking their resentments with them. The State Department had always shown itself more willing than the hardliners in the military to keep up some relationship with New Zealand, and might be encouraged in this by a lessening of the tension between the two superpowers.

All sorts of possibilities opened up in front of the ambassador, if only the magic of political contact could be worked and New Zealand moved up from its lowly ranking on the State Department's list of friends. There might be talks about economic issues. New Zealand might once again be invited to American intelligence briefings. The soldiers who went on American training courses might get the price reduced. There was no chance at all, the ambassador thought, of New Zealand and American military forces exercising together under the flag of ANZUS. But we might be able to take part in multilateral exercises in which the Americans were also taking part. The womb opened and our diplomats tried to climb right back into it.

The Americans made it clear that the prizes they held out in front of us were to be enjoyed in silence. Whatever happened, said the State Department, New Zealand was not to say that it was 'business as usual' with the United States. We couldn't be held up as an example to other wayward allies if we as much as hinted that the administration was softening its stance towards us.

In October 1988 I went once again to the United Nations in New York. The Ministry of Foreign Affairs was beside itself; its officers were terrified that I would say something in public to suggest that New Zealand was looking forward to a warming of our political relationship with the United States, and so force the State Department into denial of it. The ministry should not have worried. Politically, it did me no good to cuddle up to the Americans. Personally, I was indifferent to a meeting with whichever Republican might replace George Shultz at the State Department. Privately, I found tiresome the ministry's endless looking backwards.

I was on breakfast television in New York when a journalist asked me about the relationship between New Zealand and the United States since the ANZUS rift. I couldn't resist. 'I don't get invited to the White House. I've been to Disneyland four times and New York about three times. I've never been to the White House, and all sorts of hoods have been to the White House in that time'.

When I got home I expanded on the point. I ran through the list of criminal charges pending against senior members, or former members, of the Reagan administration. I noted its patronage of murderous generals in Latin America. The reaction was predictable. Poor Tim Francis was summoned to the State

Department, where he was told that 'moving in a positive direction' was made more difficult because of what I'd said. He put up a good rebuttal. The American ambassador in Wellington wrote to me with a personal complaint. I thought that it would probably take more than a new president in Washington to thaw my personal diplomatic freeze.

As planning in Washington for the transition to a new administration gathered speed, our diplomats redoubled their efforts to secure a resumption of high-level political contact. After my last encounter with Shultz, I'd been told that there wouldn't be any more meetings until New Zealand agreed to surrender the nuclear-free policy. Now the State Department was being less definitive about it. An assistant secretary of state told our ambassador that New Zealand would probably have to give some indication that it was prepared to revise its nuclear-free policy before our ministers were welcomed back to Washington. The ambassador pressed him on the point, but he couldn't work out if the assistant secretary meant that New Zealand was to agree to give up its policy, or simply agree to talk about giving up its policy. The department, said the ambassador, was teetering between these two alternatives.

Only in some respects did the Bush administration in fact prove more welcoming than its predecessor. In the endless round of conventions, conferences and forums in which diplomats and trade representatives carry on their business, New Zealand officials found their American counterparts more willing to speak to them, but on the political level, the Bush administration was no more appealing. Spurning members of the government, Washington still extended a welcome to National Party politicians. Vice-President Quayle, visiting Australia, asked the people of New Zealand to change their government's policy, hoping that the change would come sooner rather than later.

The low point for me came towards the end of 1989. The newspapers published a photograph of an Oval Office meeting between President Bush and Ratu Sir Kamisese Mara, the Prime Minister and stooge of the military government in Fiji. It was the day of the funeral of Timoci Bavadra, whose government and its nuclear-free policy were swept away by military force.

After that, I had even less reason to want to be seen in the White House.

The contest between politics and diplomacy ran right through the history of New Zealand's nuclear-free policy. The popular appeal of the policy was always emotional — its strength lay in its simplicity. It was at home and abroad a powerful symbol. As such it fitted uneasily into the subtle nuances of diplomacy. The instinct of the diplomat was to avoid the sweeping gesture lest it disturb the accustomed pattern of our international relationships. Like any bureaucracy, the diplomatic establishment could be moved by revolution. The refusal of the *Buchanan* visit and all that followed it was a revolution in our foreign policy, and our diplomats had no choice but to accept it. When the revolution was over, the bureaucracy started to claw its way back. The continuing tension between political aims and diplomatic aims meant that our foreign policy, and our nuclear-free policy, were not always as effective as perhaps they could have been.

Generally speaking, diplomats were happier when politicians kept their sticky fingers out of their country's international relations. They did, of course, tell politicians that contact with political leaders in other countries was very important. Politicians often believed them, because it was flattering, and because it led to meetings with the great and famous, and photo opportunities. The truth of it was that with a few rare exceptions, most international political contacts were formalities, mere showpieces. They put the seal of approval on work done in advance by the country's professional negotiators. Sometimes, meetings of political leaders could work a kind of magic which gave impetus to the negotiating process (the prospect of a superpower summit meeting was a powerful incentive to keep talking), but the leaders weren't cutting any deals when they finally sat down together. It was all worked out for them. Large international conferences were totally stage-managed.

I can only recall one or two times, at meetings of the South Pacific Forum, when political leaders actually took a decision that was different from the position their officials had worked out in advance. The officials weren't happy about it.

The most unsatisfying international gatherings I ever took

part in were the Commonwealth heads of government meetings. At the Bahamas meeting in 1985 I whiled away the time until we issued the communiqué the officials had already written by watching videos of the Prime Minister's speeches. This was all that was provided for our entertainment. The Prime Minister of the Bahamas was a remarkable character who came unscathed through an inquiry as to why in the past year he had put in his bank account an amount eighteen times greater than the total of his salary. At Vancouver in 1987 the ritual was just as empty. Here, while we waited, the delegation from Botswana ran up $1,300 worth of international phone calls and charged them to my hotel bill. The Ugandan delegation took advantage of their leader's absence at a retreat to invite a fair number of Vancouver's prostitutes to their hotel. They refused to pay and had the police evict the women. Those were the greatest excitements of the conference.

Experiences like that convinced me that international diplomacy could jog on very nicely without politicians getting involved in it. I never went to Japan as Prime Minister, but that made absolutely no difference to day-to-day business between Japan and New Zealand. Bob Hawke had lots of pictures of himself at the White House to put on his mantelpiece, but the Australians got no more trade concessions out of the United States than we did. New Zealand's diplomats didn't want a renewal of top-level political contact between New Zealand and the United States because they saw it as having value in itself; it was a symbol more than anything else. It meant that the United States had finally welcomed New Zealand back into the warmth of the old cosy relationship.

Left to themselves, our diplomats would certainly have surrendered the nuclear-free policy. Their perspective was the perspective of the State Department, Whitehall, and every other foreign ministry whose government counted itself part of the Western alliance. The test of membership of the alliance was belief in the doctrine of nuclear deterrence. As New Zealand found out, there wasn't any other test. Being a democracy wasn't enough; being well disposed towards NATO and the United States wasn't enough. You had to subscribe to deterrence to be in the alliance, and to prove it, you had to share in its risks.

In face of the undoubted fact of New Zealand's rejection of deterrence, our diplomats struggled constantly to convince the United States and its allies that we had not abandoned the values of the West. I was in the beginning an enthusiast for this approach. I knew when the Labour Party became the government that public opinion in New Zealand would not tolerate a sudden departure from the ANZUS alliance, nor would it stand for our being set apart from the countries we had traditionally regarded as our friends. Then, too, I had constant warning from my professional advisers about the price we would pay if we went too far in arousing American anger. It was not in our interest to have the Western world convince itself that we had lurched into non-alignment or anything more suspect. To meet the need, the Ministry of Foreign Affairs invented, and I adopted, a concept it called 'pro-Western regionalism'. In this scheme of things, we would serve Western interests in the South Pacific by any means short of welcoming nuclear vessels.

The result was that we locked ourselves into contradiction. Diplomats saw their task as one of isolating the nuclear-free policy and downplaying its importance. Early in the government's first term of office, I asked the Ministry of Foreign Affairs for advice on ways in which we could strengthen our position in voting on disarmament at the United Nations. This is what the Ministry had to say about NATO's refusal to disavow the first use of nuclear weapons.

'On the no first use issue, it would be more difficult for us to support a resolution. For the last three decades, the Western alliance has had its eggs in the nuclear basket . . . NATO countries also point out that they have renounced the first use of *force* and would resort to nuclear weapons only in the event of a massive Warsaw Pact invasion of Europe. In these circumstances, we could expect to run into serious difficulty with other Western countries if we were to change New Zealand policy on resolutions in this area in the General Assembly.'

This advice I accepted. Late in 1987, New Zealand voted against a resolution in the United Nations General Assembly which stated that non-first use undertakings helped reduce the danger of nuclear war and called upon the nuclear powers to assume a legal obligation not to be the first to use nuclear weapons. New Zealand's representative explained our vote by saying that the resolution looked at nuclear weapons in isolation and did not address the need for massive reductions in conventional weapons (this was NATO's inevitable justification for its policy). Nuclear deterrence, said our representative, had played a central role in security arrangements since the Second World War.

In the meantime, forgetful of those cautions, I went to a Labour Party conference and said what I actually thought about NATO and its doctrines. I stated that it was outrageous that the defence of Western Europe was based on NATO's promise to blow up the world if Russia attacked with overwhelming conventional force. NATO, I went on, had no right to decide the fate of all the rest of us. The diplomatic dovecote fluttered. The Western alliance must have thought we were schizophrenic.

Conflicting signals reflected the underlying inconsistency of our position. We said we were pro-Western, but by the Western alliance's own definition, we couldn't be. By the West's own test, we had left the alliance, but we said we hadn't. We banned nuclear weapons, then, like Russell Marshall at Geneva, we said that deterrence was keeping the peace globally. Because we decided that we must actively identify ourselves with 'Western' interests, we ended up in a kind of international halfway house. Our diplomats went round the world trying to hide the nuclear-free policy. Being seen to be inconsistent was only one result. A constant posture of diplomatic apology did nothing for our standing internationally. We were too often half-hearted in disarmament forums; having challenged the assumptions of deterrence through our actions, we tolerated it in our words.

We were still only halfway to developing a policy for our own security which was a fully fledged alternative to the nuclear alliance. While our operational alliance with the United States was a thing of the past, we were still formally allied to a superpower which had made it clear to us that its only abiding interest in the association was in the projection of its nuclear capability. New Zealand, a country which saw a nuclear defence as wrong and dangerous, and based on a mistaken analysis of the risk to its security, remained in formal alliance with Australia, a country which in that respect had a fundamentally different strategic outlook. Because the ANZUS alliance had come to be a symbol of the many interests New Zealand genuinely had in common with Australia, we remained attached to it.

More than anything else, it was our membership of ANZUS that led to the inconsistency in our policy. While we remained a member, it lured conservatives among the ranks of diplomats and politicians into thinking that it held the key to a return to the comfortable relationship we once enjoyed with the United States. It distorted our perspective with the charms of dependence. It led us too often into appeasement of deterrence and caused us too frequently to neglect our real interests. It offered nothing to New Zealand that was actually worth having. It was fool's gold. The nuclear-free policy deserved a better setting.

15

Politics and Beyond

THE LAST CHAPTER in the story of nuclear-free New Zealand is still not written. New Zealand will not take the lead it should take in nuclear disarmament until it frees itself from the dead weight of the ANZUS alliance. I had a go at putting an end to the alliance, and it just about put an end to me.

More than two years after the Secretary of State told me in Manila that the life had gone out of the defence agreement between the United States and New Zealand, the corpse was still unburied. New Zealand was still holding itself out to be a member of a three-way defence alliance with Australia and the United States. In the real world, our armed forces got to work out only with Australia. Among the politicians, the sniping continued. Representatives of the United States looked forward to the day when New Zealand would surrender its nuclear-free policy and return to the alliance. From time to time there were ripples of displeasure in the Pentagon and State Department when I talked about the pointlessness of New Zealand's inactive membership of a nuclear-armed defence alliance. Our two allies wouldn't let us in for the annual get-together of foreign ministers. In this sterile stand-off, there was little to be gained and a lot to be lost.

Meanwhile, the political situation at home was unsettled. Early in 1989 the Labour Government was hardly recognisable as the force that carried the country in the 1987 election. Our parliamentary opposition proving totally ineffective and divided, the government was filling the void itself. The disputes about economic policy, barely a flicker at the time of the election, now preoccupied us totally. The parliamentary caucus was hopelessly split into factions, the power of each depending on the shifting interests of the ambitious and the opportunist. The right-wing faction, shut out from the greatest spoils at the end of 1988, was carrying on a kind of guerrilla warfare against the rest of the government. Long-legged young women greeted the public at rallies held to provide a platform for the prime ministerial ambitions of the government's former Finance Minister. Sharebrokers and merchant bankers applauded. The rest of the country looked on, puzzled.

Against a background of recession and high unemployment, the public, seeing a government divided against itself, marked us down heavily in the opinion polls. Each poor poll encouraged the right wing to renew its efforts. Every bad result further deepened the divisions in the government. The Cabinet was paralysed by indecision, trapped by its identification with policies once enthusiastically adopted, now seen to be failing. Each meeting was riven with dispute as to who was responsible for the ignominy we found ourselves in.

Cabinet had little interest in foreign policy unless some aspect of it was a subject of domestic controversy. What was making the antennae twitch now was the proposal to buy Australian-made frigates. Their likely purchase had become a sensitive political issue, as the pressure groups which had been so supportive of the nuclear-free policy took the view, and expressed it robustly, that the purchase would lead New Zealand straight back into an active security relationship with the United States.

This was nonsense, but it was catching. I made little headway by pointing out that the ships were not going to be able to keep pace with anything the United States had in its navy. Then it occurred to me that if New Zealand could formally disengage itself from its alliance with the United States, the argument from the pressure groups would lose its force.

With this crudely political motive in mind, I raised with Kim Beazley the possibility of New Zealand's ending its formal military association with the United States. The Australian Minister of Defence had come to Wellington to talk over the frigate purchase with New Zealand ministers, and the peace groups marked his arrival by plastering on every lamp-post in the town his bare-chested image in the fashion of Stallone. In person, his suit jacket shrouded an elderly woollen pullover oddly out of harmony with the rest of him. I met him in the company of our new Minister of Finance, small and nervously energetic. The meeting had no formal agenda. I floated past Beazley the idea of our calling it quits; there was no visible reaction. We had no chance to talk about it at any length and I left the meeting intending to think about it further.

One obvious problem lay in the wording of the ANZUS treaty itself. On its face, it continued in force indefinitely. The United States had got around this difficulty by announcing, in effect, that it no longer considered itself bound by the treaty as far as New Zealand was concerned. This action having no basis in international law, it was not a precedent I wished to follow. The only part of the treaty that hinted at its ceasing to function was a clause allowing each member of the alliance to give twelve months' notice of withdrawal from the ANZUS council. The council not having met for four years, our withdrawal could only be symbolic. But it was as much as we could do, and would effectively put an end to our part in the treaty.

Given its government's wish for New Zealand to abandon its nuclear-free policy, I was certain that Australia would cry foul if New Zealand gave notice of withdrawal from the ANZUS council. To frighten us back into the fold, the accusation would certainly be made that New Zealand was risking its military association with Australia. I thought we could live with the noise. There wasn't anything altruistic in Australia's military involvement with New Zealand; what we got from them we paid for. We worked together because it suited us both. Inside ANZUS or outside, the co-operation would continue.

In terms of our domestic politics, American displeasure could be handled. By now it was plain to everyone that there would be no more co-operation between New Zealand and the conventionally armed forces of the United States. The price of the resumption of military activity was our surrender of our nuclear-free policy, a price most of the public were now unwilling to pay. The spectre of imminent peril which once loomed outside the sheltering bounds of ANZUS had simply vanished. The country had weaned itself from the idea that we were inevitably a dependent of the United States; we had effectively cut the cord without doing ourselves any substantial injury. In fact, our assertion of independence had lifted our spirits as a country. Our nuclear-free policy was becoming part of our national identity.

What started out as a political gambit became something more carefully calculated. Leaving ANZUS was a necessary step towards the resolution of the inconsistency in our foreign policy. We would never again enjoy our cosily dependent relationship with the United States, but our view of the superpower would inevitably become more realistic. Once the Americans got their immediate outrage off their chest, withdrawal from the ANZUS council would take the contradictions out of the relationship. As long as we stayed an ally, in name if not in fact, we couldn't build an honest relationship with the United States. While we were in the alliance, the basis of our relationship could never be plain dealing; it would rest on concessions secretly sought and sparingly given. Approaching our erstwhile ally always as a supplicant, we tainted our nuclear-free policy and debased our independence.

The United States looked at it from the other end of the telescope and came up with the same result. A deputy assistant secretary of state, seen by our ambassador in Washington in his ceaseless search for a warming of the relationship, put it like this. The administration intended to work constructively with New Zealand in areas where this was natural and possible, like multilateral trade and economic policy. But this could not obscure the point that there were what he called persistent and unresolved alliance issues between us. They were, he said, 'like a mote in the eye', and whenever the United States looked at the relationship, 'we will always be conscious of them as an abnormality'. With this diagnosis, I could only agree, however much we might differ about the remedy.

To the fractious Cabinet I put the possibilities. I discussed with them the difficulties of the political management of the frigate purchase. I suggested that some speaking engagements I had accepted in the United States at the end of April 1989 might be turned to our advantage if I raised the prospect of New Zealand's giving notice from the ANZUS council. If any minister had any instinctive reaction to the proposal, it remained unspoken.

Another Cabinet meeting followed. What we were discussing I have long forgotten, when suddenly a minister, schoolboyish in appearance and a favourite tool of the right-wing faction, asked if I would be giving notice of withdrawal from the ANZUS council when I went to the United States. (This minister had not before been known to take an interest in our foreign policy.) Sensing a trap in the making, but not yet seeing its form, I was terse in my response to him. I pointed out that it was impossible for me singlehandedly to give notice of withdrawal from the ANZUS council. That required a formal decision of Cabinet, which must agree to take the formal procedures needed for an amendment to an international obligation. It was the end of the discussion.

When the issue was once again raised at another Cabinet meeting just before I left for the United States I said very simply that my major speech in America would be dealing with ANZUS and what might happen to the alliance. I repeated that I could not and would not be giving formal notice of withdrawal from the ANZUS council. Some discussion then took place on the merits of withdrawal. Most of those who spoke took the view that withdrawal should not be pursued lest there be some disturbance in an already strained relationship with the United States. The minister responsible for trade expressed concern about the understandings we had already reached with the Americans, essentially amounting to under-the-table discussions between senior officials about trade issues. These could be put at risk by any unfavourable American reaction to our giving notice.

The debate petered out. It was perfectly plain to Cabinet that the question of withdrawal was yet to be settled. All were agreed that it would take careful and prolonged deliberation before it was settled.

In the meantime John Henderson, head of the Prime Minister's Office, had been taking soundings of the likely international response to New Zealand's withdrawal. It was his responsibility to inform me of the views of the Australian Government. He also kept in constant touch with the American embassy so that I could be aware of whatever undercurrents were to be found there. Henderson was a careful operator. He told me that the response from both countries would be much more than unhelpful. In his view, both would put pressure on New Zealand great enough for the issue to become a political liability.

John Henderson's faithful reporting did not persuade me. Our freedom of action was constantly subverted by our lingering attachment to a meaningless alliance which served none of our interests. We did the defence business we needed to with Australia under other agreements. Yet there we were still, locked into the past, threatened always with the displeasure of our nominal allies if our actions questioned the assumptions that underpinned the alliance. It was time to lance this particular boil. I set off for the United States determined to put the prospect of New Zealand's giving notice of withdrawal from the ANZUS council into the public arena.

Once in New York, I found reminders of the extraordinary tolerance and diversity of the United States. I went to the Dakota Building opposite Central Park. Yoko Ono had planted a tree for New Zealand in the grove in the park she endowed in her husband's memory; she talked to me about nuclear and environmental issues in the chilly April dusk. She was one among many in the United States who had written to express their support for New Zealand's nuclear-free policy. The next day I spoke at an Anzac Day commemoration. On the fifth floor of the Rockefeller Centre was the Anzac garden, where each year Australians and New Zealanders joined American veterans to honour the dead. There were in that setting only echoes of the Anzac ceremonies of my childhood in Otahuhu, but the unifying spirit of the day was as strong as it had ever been.

I had a speaking engagement at Yale; I had been asked to deliver the George Herbert Walker memorial lecture, endowed by the uncle of the president. I liked the irony. The people I met at the venerable university were welcoming, liberal in their outlook and generous in their response to a speech which however mildly, and, I hoped, fairly put, was in the end a substantial criticism of the policy of the United States Government. I spoke of the many common interests New Zealand shared with the United States. I told my audience of the disappointment I felt that the American administration was reluctant to further those interests because of a disagreement about nuclear disarmament. I made clear my view that it would be better for the two countries to agree to disagree and put the dispute behind us. The security alliance between the United States and New Zealand was a dead letter. The basis of the ANZUS alliance was a commitment between the treaty partners to consult. New Zealand was excluded from ANZUS council meetings and there was no longer any consultation. That raised the issue of whether New Zealand should give formal notice of withdrawal from the council.

I knew that this carefully worded speculation would cause howls of outrage in Canberra and Washington. I looked forward to it. I believed from past experience that I could turn their disapproval, with its inevitable bullying overtones, into advantage for the government. If the government's political opponents were provoked into a formal reaffirmation of their commitment to ANZUS, and so, as all New Zealand understood, to acceptance of nuclear weapons, all the better for the government. All we had to do was keep our nerve.

The clocks in New Zealand are eighteen hours ahead of those on the east coast of the United States. When I spoke at Yale the time at home was early in the morning of 25 April. All over the country folk were getting ready for Anzac Day services. Now the announcement was suddenly made to them that New Zealand was withdrawing from ANZUS. The subtleties of what I said went by the board. The possibilities I had so carefully canvassed were disregarded. The reporting simply carried the message that we were leaving. On the day set aside to remember the suffering of war, ANZUS was once again a powerful totem. To question its magic was to question the sacrifices the day represented.

My parliamentary colleagues, arrayed in attendance at services of commemoration throughout the country, found themselves molested by angry representatives of the most conservative elements in the community. They were entitled, given the nature of our discussions in Cabinet, to be astonished at what must have seemed from the early reporting to be a reckless and improper act on my part. In the circumstances, they were inevitably disconcerted, their assurance shattered. This was fatal. My colleagues' uncertainty was soon communicated to the public, which, feeding on it, magnified their doubts a hundredfold.

I was never able to recover the ground. I lost my temper in my anger at the reporting of my speech. I might as well and to as much effect have shouted at a storm cloud. In a lobby of the Canadian House of Commons I sat on what looked like a bar stool and spoke into a camera to an audience thousands of miles away in New Zealand. Curious passers-by looked on. My outburst made me feel better, but in the end it did me no good. No amount of repetition of what I had actually said could take away the sting of Anzac Day morning.

Knowing that many of my colleagues in the government were trying to distance themselves from the proposal by disavowing any knowledge of it, the Australians and the Americans complained that they too had not been consulted. Having long before been told that we were considering, amongst our options, the possibility of withdrawal from the ANZUS council, they can only have meant that I was not to mention it without their agreement.

Their complaints in any other circumstances would have been of little moment, and easily turned against them. Now they added to my difficulties. Kim Beazley played on the Anzac chord. The ANZUS treaty was not a dead letter, he claimed, blithely misrepresenting what I said about the undeniably defunct arrangement between New Zealand and the United States. It was still the basis of the defence relationship between Australia and New Zealand. This was so obviously untrue, and an assault so much expected, that I would have enjoyed debating it. But now it was too late.

Factional plotting finished it off. Two members of the right-wing clique in the Cabinet took it on themselves to whisper to journalists that I had, in making my speech at Yale, breached an agreement of the Cabinet. This was plainly not true, but the disarray among the government was fertile ground in which the lie was planted. Whatever political capital I might once have had was now played out.

This débâcle meant an end to any immediate chance of our putting our foreign policy on a solid footing. New Zealand could not now pursue the option of giving notice of withdrawal from the ANZUS council. The views of Australia and the United States were immaterial. It was the New Zealand response that made the course impossible; the domestic debate on the issue had been totally corrupted. The issue had been transformed into a question of whether we were provocatively insulting former allies with whom our soldiers had fought and died. No government could win a fight on such a question unless it was firm in the courage of its convictions and united in pursuit of its goal, but of that there was no prospect. The government, once easily rallied by the call to defend our nuclear-free policy, would rally no more. No policy was more compelling than the cry of faction.

Foreign policy now spilled over into domestic politics. I found in the distance between New Zealand and North America more than physical isolation from my colleagues in Cabinet. I was disappointed in the failure of Deputy Prime Minister Geoffrey Palmer to support in public my account of what had taken place in the Cabinet. When I got back home, I found out what had happened. Palmer met me at the airport in a state of high excitement. 'You haven't got the numbers!' was his greeting to me.

Palmer was the only serious contender for my job as leader of the parliamentary Labour Party. I had been sustained through many difficulties by his unfailing loyalty; in the factional warfare that was destroying the government, many chances had been given to him to unseat me, but he had taken none of them. I believed him when he told me, as he did from time to time, that if I were not the leader of the party, he would not stay in politics. Now he had changed his mind. He would not challenge me as leader of the party, he said, but if I no longer held the place, he would stand for election to it.

That, I knew, was the end of it. Palmer's refusal to make himself available for election was my one sure safeguard. What made him move, I cannot say. Whatever his reasons, an argument about ANZUS had given him his opportunity. I went from the airport to the campus where the Labour Party was holding its Auckland regional conference. The delegates swept me up in warm endorsement and for a while I forgot my difficulties. The next day fist-fighting broke out among the factions. Peace seemed further away from the Labour Party than ever.

On the Monday morning before the Cabinet met, Palmer gave voice to a doleful monologue. There was dissatisfaction among the Cabinet. There was stirring among ministers, he told me. At Cabinet itself, there was the usual chorus from the aggrieved and the aspiring. There was the usual firing from entrenched positions. Nothing was decided; it was the dismal paralysis of a divided government.

People sometimes ask me if the CIA had anything to do with my resignation from office in August 1989. The fact is that it didn't. Or if it did, its finesse was such that nobody has ever heard of it, or suspected it. The only dirty tricks that were ever played on me were played by right-wing members of the Labour Party. Neither the CIA nor any force on earth could tear us apart the way the Labour Government tore itself apart. In the end, that was all there was to it.

My last dealings with the United States administration were as unsatisfactory as all the rest. When I was in Ottawa, Canadian Prime Minister Brian Mulroney told me that the American view of New Zealand was short-sighted. The United States was misguided to cut itself off from high-level political contact with New Zealand while welcoming others with dubious credentials. He offered to raise the matter at his next meeting with President Bush, which he did, asking the President if any thought had been given to resuming contact with New Zealand at the senior political level. No, said the President, the New Zealand Prime Minister's speech at Yale had set the relationship back. The message from the White House could hardly be clearer. Any New Zealand minister who wanted the welcome mat from official Washington must first, whatever other price may be exacted, keep silent about New Zealand's nuclear-free policy.

Then the climate altered. Six months after my resignation, Secretary of State James Baker announced that he would resume political contact. He broke the ice by meeting the New Zealand Minister of Foreign Affairs in Washington. The encounter our diplomatic establishment had for so long sought had finally taken place.

I thought that in public I should make the best of it. Publicly, I was glad that the United States had finally come to its senses and had put the relationship back on an adult footing. Privately, I was less confident. I did not know the price of the meeting, but what I did know about our dealings with the United States suggested that, one way or another, some toll had been paid. Perhaps it was our shameful vote in the United Nations in support of the American invasion of Panama, a vote that contradicted all we had ever argued about the right of small nations to the protection of international law. Perhaps we had reached an understanding that we should unfailingly be advocates of the American view of deterrence. Whatever might have been the government's intentions, we were once again ensnared in the contradictions of the nuclear alliance.

The gloom which the Baker meeting induced in me was soon dispelled, and from a most unlikely source. The National Party suddenly announced that it would adopt the nuclear-free policy of the Labour Government. Its leader said that his party, when it became the government, would not repeal the law excluding nuclear weapons from New Zealand. The National Party, in effect, would no longer rely on the United States to respect New Zealand's nuclear-free policy. Whatever the result of the next general election, there would be no nuclear warships in New Zealand ports.

Political commentators, agog at this reversal, looked for an explanation for it. The National Party itself cited the changes taking place in Eastern Europe, and said that its alteration in policy was the appropriate response. Some observers thought that the party's thunder had been stolen by the Baker meeting. The National Party had always boasted of its ability to restore the political relationship between New Zealand and the United States, but here was the relationship restored without it. Still others put the party's decision down to a simple political calculation. If it changed its policy, it would take away from the government the most appealing and distinctive policy the government had to offer the voters.

I think that it came down to political reality. Nuclear-free New Zealand has gone beyond politics; support for the policy had long since detached itself from opinion about the government. I don't like to recall the depths we plumbed in the polls when support for the ban on nuclear ships was recorded among more than eighty per cent of the electorate. This is something I don't see altering. There is now engrained in the New Zealand public a conviction that New Zealand was right to deny access to nuclear ships and right to stand aside from the nuclear arms race. Nuclear-free New Zealand cannot be touched by the accidents of politics or the wishful thinking of our one-time allies. However much politicians and diplomats may try to pretend that the policy doesn't exist, the fact is that it does. It's bigger than all of us, the solid rock that cannot be moved. New Zealand will always be an affront to the advocates of deterrence, and always be a comfort to the supporters of disarmament. And that, in the end, is all we set out to be.

POSTSCRIPT

New Zealand had to fight to become nuclear-free. There was never a day in the five years I was Prime Minister when I had the luxury of dropping my guard. It happened like that because New Zealand was a member of one of two implacably confrontational power blocs. The United States was the acknowledged leader of NATO and ANZUS. To round off the global nature of its quest for supremacy it projected its power singlehandedly in areas such as the Indian Ocean where there was no treaty cover. Given its strategy and enormous investment it could brook no dissent in the ranks. New Zealand's duty was to be uncomplainingly swept up in exactly the kind of international totalitarianism we were supposed to be ready to defend ourselves against. The rigour of this collective discipline was so important that the gross violation of New Zealand's sovereignty by France when it bombed the *Rainbow Warrior* and killed a man to avoid international publicity over its nuclear-testing programme went uncondemned and unremarked by the leaders of the Free World.

The nuclear crescendo has now been called into question. Not by coolheaded self-protective analysis but by the remarkable events in Eastern Europe and the Soviet Union. The Soviet bloc has broken up in front of us and the Soviet Union itself is in jeopardy from within. The assumptions of threat, which have not been challenged for forty years, are going to be tested. The Soviet Union will probably remain in the ranks of great powers but it will not be a superpower. There will only be one superpower — the United States.

The assumptions will be tested in the Northern Hemisphere. The doctrine of nuclear deterrence as it is practised by the member states of NATO rests on the assumption that there is an overwhelming threat to Western Europe from the conventional forces of the Soviet Union and its satellites. Those massive concentrations poised against Western Europe are deterred from invasion by the announced intention of NATO to use nuclear weapons to repel the invasion. This posture fails to meet the test of its own internal logic if the threat of invasion from the East reduces. If East Germany takes itself (and its MIG fighters!) into NATO through the conduit of West Germany, NATO will have increasing difficulty persuading Western Europeans that it is incapable of matching the Soviet Union from its own resources of conventional armaments.

As the Soviet bloc continues its fragmentation countries will come to make their own judgments about the threat to them and the appropriate response to it. There will be domestic pressures for reductions in defence spending. There will be less concern about the eruption of nuclear war. There will certainly be a lessening in the solidarity of the Western alliance. People do not willingly bear the burden of a nuclear defence with its awesome risks and horrendous economic cost unless they are very frightened. If they are not frightened they won't put up with it. The difficulties that some European governments faced in 1986 when they persuaded their populations to accept deployment of Cruise missiles were overcome. They wouldn't be today. It is simply not possible to argue, credibly, that world communism is a single threat to which the West must always make a single co-ordinated response.

Over the last five years New Zealand has been not so much out-of-step as ahead of its time. We made our assessment of risk just as Western Europe will do now. We decided it was pointless to arm ourselves with a nuclear defence. We did not see the disturbances which take place from time to time in the South Pacific as the work of communism in search of world domination. We chose not to embrace the risks that accompany nuclear deterrence wherever it operates in the world. To go nuclear in the South Pacific is to invite greater instability.

That kind of assessment will become more commonplace as the threat from the Soviet bloc recedes. It would be remarkable if many Americans did not look at the world from a new perspective. In the 1960s it would have been difficult to see Cuba as anything other than the near outpost of an alien power. In the 1990s it may be possible to see left-wing political movements in the Caribbean and Latin America not as the tool of Soviet communism but as an indigenous response to local history and circumstances. Threats will upset the stability of the region but the nature of the threats will be seen as fundamentally different.

None of this means that the world has become safe. My instinct is to rejoice in the collapse of the totalitarian regimes of Eastern Europe. Their going is a victory for the human spirit. But the irony of the triumph of democracy over vassallage is that the United States and the Soviet Union can't cut deals any more and coerce their implementation. The breakaway states will go through the agonies of radical economic and social change, which will bear hard on their citizens, and survival as democracies cannot be assumed. Their assertion of independence seems non-negotiable.

There are no certainties in the quest for security. Once the balance of terror gets terminally out of kilter other mechanisms will be sought to bring order to the affairs of nations. Other analyses will have to be tendered to explain why we are all still here. There seems, from a small state perspective, to be only one hope, which is that the United Nations will be mandated by the great powers with the authority to settle issues that arise between nations. That would involve a loss of a sovereign right to pursue national interests to the exclusion of other nations' interests. It would reverse the practice of forty-five years whereby great powers cut deals and the United Nations became the patron saint of the settlements. It would involve a whole new body of international law and a radical approach to the enforcement of adjudications. It would require true greatness from the powers whose consent is required. But in the imbalance which is now the fact both the Soviet Union and the United States would have much to gain from the more rational conduct of international relations which that submission would produce.

Small powers cannot wait until the big powers move. They will assess their security options. They will take slow and often painful steps to minimise their risks. The exclusion of nuclear weapons from New Zealand was a small step to take in the global context, but unless a lot of small steps are taken we will always be in thrall to nuclear weapons.

MORE ABOUT PENGUINS

For further information about books available from Penguin please write to the following:

In New Zealand: For a complete list of books available from Penguin in New Zealand write to the Marketing Department, Penguin Books (N.Z.) Ltd, Private Bag, Takapuna, Auckland.

In Australia: For a complete list of books available from Penguin in Australia write to the Marketing Department, Penguin Books Australia. P.O. Box 257, Ringwood, Victoria 3134.

In Britain: For a complete list of books available from Penguin in Britain write to Dept EP, Penguin Books Ltd, Harmondsworth, Middlesex UB7 0DA.

In the U.S.A.: For a complete list of books available from Penguin in the United States write to Dept DG, Penguin Books, 299 Murray Hill Parkway, East Rutherford, New Jersey 07073.

In Canada: For a complete list of books available from Penguin in Canada write to Penguin Books Canada Ltd, 2801 John Street, Markham, Ontario L3R 1B4.